INTRODUCTION TO ESTUARINE HYDRODYNAMICS

This textbook provides an in-depth overview of the hydrodynamics of estuaries and semi-enclosed bodies of water. It begins by describing the typical classification of estuaries, followed by a presentation of the quantitative tools needed to study these basins: conservation of mass, salt, heat, momentum, and the thermodynamic equation of seawater. Further topics explore tides in homogeneous basins, including shallow water tides and tidal residual flows, wind-driven flows in homogeneous basins, density-driven flows, as well as interactions among tides, winds and density gradients. The book proposes a classification of semi-enclosed basins that is based on dominant dynamics, comparing forcing agents and restorative or balancing forces. *Introduction to Estuarine Hydrodynamics* provides an introduction for advanced students and researchers across a range of disciplines - Earth science, environmental science, biology, chemistry, geology, hydrology, physics - related to the study of estuarine systems.

Arnoldo Valle-Levinson is a professor in the Department of Civil and Coastal Engineering at the University of Florida, Gainesville. Throughout his career, he has carried out observational and theoretical studies on exchange processes in semi-enclosed ocean basins. His work has concentrated on the hydrodynamics of fjords and estuaries, as well as of temperate, subtropical, and tropical systems. He is an editor for *Continental Shelf Research*. He is an associate editor for *Estuaries and Coasts* and the *Journal of Oceanography*. Valle-Levinson is the editor of the book *Contemporary Issues in Estuarine Physics* (2010, Cambridge). He has published over 200 peer-reviewed articles and is a Corresponding Member of the Mexican Academy of Sciences. An estuary in Chile has been named in his honor: ‘estero Arnoldo.’

INTRODUCTION TO ESTUARINE HYDRODYNAMICS

ARNOLDO VALLE-LEVINSON
University of Florida

CAMBRIDGE
UNIVERSITY PRESS

CAMBRIDGE
UNIVERSITY PRESS

University Printing House, Cambridge CB2 8BS, United Kingdom

One Liberty Plaza, 20th Floor, New York, NY 10006, USA

477 Williamstown Road, Port Melbourne, VIC 3207, Australia

314–321, 3rd Floor, Plot 3, Splendor Forum, Jasola District Centre, New Delhi – 110025, India

103 Penang Road, #05–06/07, Visioncrest Commercial, Singapore 238467

Cambridge University Press is part of the University of Cambridge.

It furthers the University's mission by disseminating knowledge in the pursuit of education, learning, and research at the highest international levels of excellence.

www.cambridge.org
Information on this title: www.cambridge.org/9781108838252
DOI: 10.1017/9781108974240

First published 2022

A catalogue record for this publication is available from the British Library.

Library of Congress Cataloging-in-Publication Data

Names: Valle-Levinson, A. (Arnoldo), author.
Title: Introduction to estuarine hydrodynamics / Arnoldo Valle-Levinson.
Description: Cambridge, United Kingdom ; New York, NY : Cambridge University Press, 2022. | Includes bibliographical references and index.
Identifiers: LCCN 2021037808 (print) | LCCN 2021037809 (ebook) | ISBN 9781108838252 (hardback) | ISBN 9781108974240 (epub)
Subjects: LCSH: Estuaries–Hydrodynamics.
Classification: LCC GC97 .V35 2022 (print) | LCC GC97 (ebook) | DDC 551.46/18–dc23
LC record available at https://lccn.loc.gov/2021037808
LC ebook record available at https://lccn.loc.gov/2021037809

ISBN 978-1-108-83825-2 Hardback

Contents

Preface

This text is the result of more than 30 short (1–2 week-long) courses and a handful of semester-long courses on estuarine hydrodynamics. Its production has been encouraged throughout the years by colleagues who have organized some of the short courses: David Salas de Leon and Alex Souza in Mexico, Guto Schettini and Eduardo Siegle in Brazil, and Aldo Sottolichio in France. In the preparation of lectures, Larry Atkinson in the USA has been a great source of inspiration and advice. I am thankful to all of them for their steady encouragement to turn course lectures into a book.

The material in the text has benefited from many colleagues with whom I have interacted throughout the years. Professional meetings and project collaborations often provide opportunities to recognize, refine, and reinforce field underpinnings. In many ways, this text is a community effort spawned from generous and collegial sharing of thoughts. Just like the saying "it takes a village to raise a child," it took the "global estuarine village" to raise this "child."

The presentation level of the material in this text intends to be for *beginner* to *intermediate* skills. To the nonspecialist, however, a rapid page perusal may cause an irascible reaction: "beginner"? "intermediate"? malarkey! I hope that is not the reaction. If indeed it is, it is probably because the perusal triggers sensory overload from many equations, despite efforts to minimize their number. To that, I will respond that equations are your friends. They actually represent lots of words in a few symbols. So they in fact save you time. In some instances, those symbolic expressions even describe a mathematical problem that has a solution. And that solution represents what happens or approximately happens in nature. So cool! My plea is, try to be patient with your friends, the equations; they are helpful.

The text sometimes may be repetitive, either in the same section or in the chapter, or from one chapter to the next. This has been done in an attempt to reinforce concepts. Some passages may present more information than needed, while others might not have enough. The intent is to present topics from an introductory approach. The focus of the material presented is by no means comprehensive, and it is extremely likely that crucial references are overlooked. Motivated readers are urged to explore literature suggested at the end of each chapter – in additional sources and the studies referenced therein.

The book follows the sequence of topics featured at the short courses mentioned in the first paragraph of this Preface. It starts with an introduction and a typical classification of semienclosed coastal basins influenced by fresh water. This is followed by the presentation of the quantitative tools to study these basins, namely conservation of mass, salt, heat, momentum, and thermodynamic equation of seawater. The first phenomenon treated is tides in homogeneous basins, including shallow water tides and tidal residual flows. The text continues with a treatment of wind-driven flows in homogeneous basins. The following topic is density-driven flows – the typical gravitational circulation. After that, the text explores the interactions among tides, winds, and density gradients. This is the longest chapter and the one with a large number of unresolved research questions. The next topic has to do with fronts, followed by times of water renewal in semienclosed basins, and behavior of basins with low-river discharge. The text ends with a proposed classification of semienclosed basins that is based on dominant dynamics, comparing forcing agents and restorative or balancing forces.

Several of these chapters benefited enormously from relevant books or chapter work by Bruce Parker (shallow water tides), Chunyan Li (tidal residuals), Clinton Winant (wind-driven flows), Jim O'Donnell (fronts), Lisa Lucas (renewal times), and John Largier (low-discharge basins).

Finally, an effort of this nature would be impossible without the inspiration from loved ones. Thank you to my beloved Anne and our wonderful Liliana, Alvaro, and Emiliano for their patience, support, and understanding.

1

Introduction and Classification

1.1 Introduction

This book is intended mainly as an *introductory* source for researchers focusing on any aspect of estuarine studies, for example, biological, biochemical, geochemical, physical. More generally, this volume is intended for those investigating all facets of semienclosed coastal bodies of water. The text presents information at a *basic level*, but perhaps moves up to an intermediate level in some sections. The book should provide fundamental concepts to interdisciplinary studies in these systems and to advanced hydrodynamics investigations. The core purpose of this text is to offer relevant concepts accessibly. In some parts, the offerings are succinct and in others they are reiterative. The reason for occasionally restating or reiterating concepts is to try to present them from different angles in search of lucidity.

Unquestionably, there is a surging need to understand the wide array of processes that occur in semienclosed coastal bodies of water, including estuaries. These systems appear sprinkled along coastlines throughout the world. Close to 75% (26 out of 35) of the largest cities in the world, in terms of population, are found in or influenced by estuarine systems. A higher percentage has a direct impact on estuarine waters through river connections. Worldwide, estuaries serve various purposes: they help preserve freshwater resources in rivers by limiting saltwater intrusion; they provide habitat and nursey grounds for commercially and ecologically pivotal species; they act as river-related sediment and pollutant receptacles, hampering contamination in coastal waters; they enable tourism and recreational activities; they host harbors and related commercial activities like fishing; and they could potentially provide clean energy. Threats from sea-level rise, from increased pollution, saltwater intrusion, freshwater consumption, and storm intensity, as well as threats from changes in river discharge, are already occurring in coastal regions. Therefore, we need to understand current conditions affecting semienclosed basins so that we can propose possible consequences to all of their functions (some of them mentioned above).

1.2 Classification

When studying the hydrodynamics of semienclosed basins, we may sometimes be distracted by trying to determine whether the basin is an estuary. According to the originally proposed definition, an estuary has three main characteristics. First, it is a semienclosed basin or body of water. Second, the basin has free communication or connection with the ocean. Third, land-derived drainage dilutes ocean salinity concentrations within the basin. Under these conditions, tidally averaged flows, or flows averaged over days, months, or years, should display buoyant outflow of estuarine waters and inflow of relatively heavier oceanic waters. This circulation is referred to as *gravitational circulation* as it results from density gradients under gravity's action. Gravitational circulation can also be called density-driven *exchange flow* as it describes mean or residual flows moving in opposite directions.

We need to remember that the concept of gravitational circulation disregards instantaneous tidal flows. We also shall consider that there are examples of semienclosed basins that display gravitational circulation with no dilution of ocean waters (e.g., Chapter 11). These basins may not fit the "formal" definition of an estuary but certainly behave dynamically as one. Moreover, there are basins that show opposite or reverse (or inverse) gravitational circulation with inflow of relatively buoyant waters and outflow of heavy (or *hyperpycnal*) basin waters. Such a condition appears in highly evaporative basins where ocean water becomes concentrated with salts. Again, just because evaporative basins fall short in fulfilling the "formal" definition of an estuary, it does not preclude them from dynamically behaving as one. Therefore, defining an *estuary* becomes a challenging proposition: difficult to define but we know when we encounter one.

Classifying estuaries or semienclosed basins in different categories may provide general information on how they behave dynamically. However, classifications should be used with caution as they overlook crucial details of each system. Nonetheless, we present here different ways of classifying estuaries on the basis of their *geomorphology*, their *water balance*, and their *water column stratification*. More comprehensive classifications are described in Chapter 12, after understanding basic physical mechanisms in estuaries.

1.2.1 Classification Based on Geomorphology

Based on their geological origin, semienclosed basins may be grouped into the following categories: *drowned river-valley* or *coastal plain*; *fjord* or *drowned glacier-valley*; *tectonic* and *bar-built* (Figure 1.1).

Drowned river-valley or coastal plain estuaries were river basins more than ~15,000 years ago, during the last glacial maximum, when mean sea levels were roughly 120 m below present day. Drowned-river valley estuaries took shape

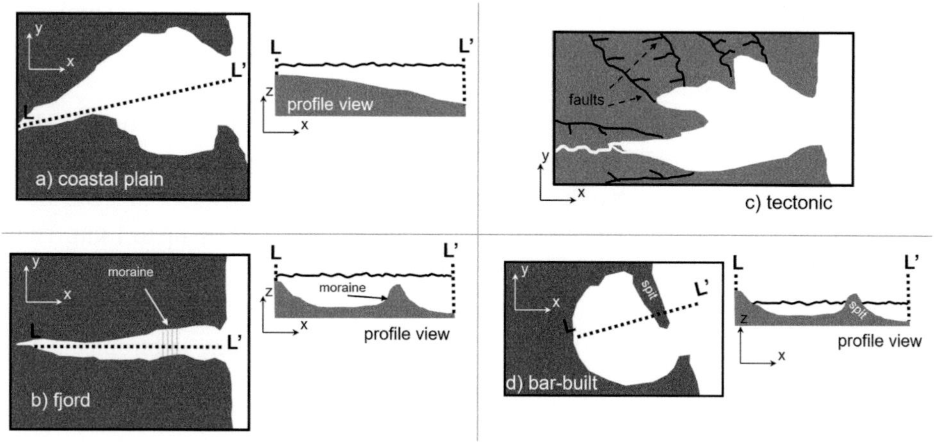

Figure 1.1 Classification based on geomorphology showing the four different types, with typical cross-sections along transect L to L′.

between 15,000 and 8,000 years ago, as sea levels rose because of ice melting, and flooded the former river basins. Typical examples in the United States include the Chesapeake Bay, Delaware Bay, and the Hudson River. River basins continue to modify these estuarine basins, which typically are influenced by comparatively heavy sediment loads and have relatively shallow depths (order 10 m).

Fjords were formed in a similar way as drowned river-valleys. Instead of having been rivers, fjords were glaciers or "rivers of ice." Fjords took shape as glaciers melted through flooding the glacier valley with relatively warmer ocean waters. These basins are relatively deep (order 100 m) and have steep-sided bathymetries. Many fjords receive freshwater input from river discharge. At the highest latitudes, however, some fjords can still be fed by glaciers (tidewater glaciers or ocean-terminating glaciers). Fjords appear in relatively high latitudes (e.g., Scandinavia, Greenland, Antarctica, British Columbia, Alaska, Chile, and New Zealand).

Fjords with tidewater glaciers provide direct information on climate change. The interaction between the fjord's gravitational circulation and the ocean-terminating glacier can represent a feedback mechanism that accelerates glaciers' melting rates. The more the glacier melts, the stronger the gravitational circulation and the larger the heat fluxes from warm ocean waters to the submerged glacier face, thus enhancing its melting. Implications of this feedback are also far-reaching as melt water from land, through its seaward advance in the glacier, can contribute to rising sea levels.

Tectonic estuaries have been formed by earthquakes or faults that suddenly or gradually shift the Earth's crust adjacent to the ocean over distances on the order of kilometers. Such motions of the crust, which can be horizontal or vertical, allow ocean incursions. In some cases, such as in the San Francisco Bay in California, tectonic faults may also allow channelization of river valleys. Other examples of tectonic semienclosed basins are found in the Galician Rias of Spain, Guaymas Bay, Mexico, and Manukau Harbor, New Zealand.

Bar-built estuaries have become semienclosed basins because of sediment drift along the coast, or littoral drift. This sediment transport elongates from a headland, giving way to a spit or sand bar that separates the mainland from the ocean. Bar-built basins are characterized by a relatively narrow inlet, of order 100 m up to 1 km. They often display several inlets that interrupt barrier islands and sand bars. Bar-built estuaries may also be formed by rising sea levels that isolate barrier islands separated by inlets. This type of estuary is widely found in temperate, subtropical, and Mediterranean climates. Examples abound in the east coast of the USA.

1.2.2 Classification Based on Water Balance

On the basis of *water balance*, semienclosed basins may be characterized in three types: *positive* or *typical* estuaries, *negative* or *inverse* estuaries, and *low-discharge* basins (Chapter 11). Positive or typical estuaries are the most widely studied. They are basins influenced by a positive water balance, that is, gains of fresh water. This means that freshwater input from river discharge plus precipitation (rain), plus ice melting and submarine groundwater discharge exceeds water losses from evaporation plus freezing and seepage through the ground. Positive estuaries display relative buoyant water inside the basin compared to the ocean's water density. Over the long term (days to months to years), the situation described by a positive density gradient (Figure 1.2a) favors the gravitational circulation that consists of outflow of buoyant waters and inflow of ocean waters. Typical estuaries are found predominantly in temperate regions or in weak-seasonality tropical climates.

Inverse estuaries are found in arid and semiarid areas where water losses prevail over water gains. This situation establishes negative density gradients that are inverse to those in typical estuaries. Inverse density gradients drive, over the long term, inflow of relatively buoyant waters and dense-water outflow (Figure 1.2b). Because of net water losses, inverse estuaries exhibit sluggish water flushing and should be more susceptible to water quality problems than typical estuaries.

Low-discharge basins are also found in arid and semiarid regions, including Mediterranean climates. In these systems, evaporation rates may be on the same order of magnitude as freshwater input rates. Depending on whether the buoyancy inputs are from heat fluxes or water fluxes, the basin's circulation may vary from typical to inverse or, in some cases, may exhibit *salt-plug* characteristics (Figure 1.2c). In the *salt-plug* situation, the landward portion of the basin displays gravitational circulation. The portion connecting to the ocean exhibits inverse

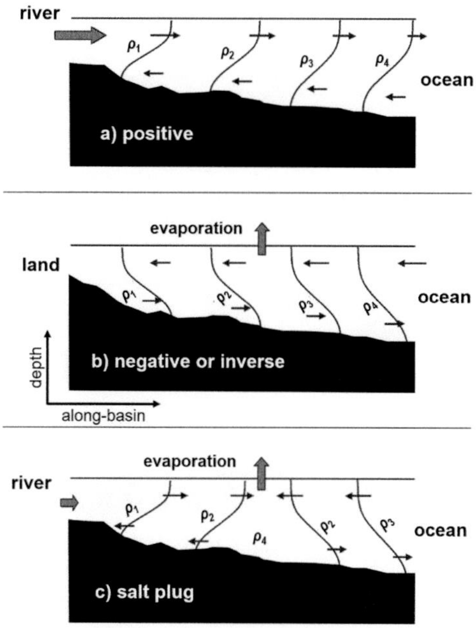

Figure 1.2 Classification based on water balance showing the three different types, with typical along-basin cross-sections of density and flow. Arrows are for illustrative purposes as they are not scaled proportionately to other frames. In all density isolines (isopycnals), $\rho_1 < \rho_2 < \rho_3 < \rho_4$.

gravitational circulation because of a region of local salinity maximum inside the basin. These low-discharge basins attract the attention of an entire chapter (Chapter 11) in this volume.

1.2.3 Classification Based on Water Column Stratification

This approach considers the qualitative competition between stratifying and mixing agents in a semienclosed basin. Stratification is typically induced by buoyancy forcing from river discharge while mixing arises mostly from tidal or wind forcing. Another way of regarding the buoyancy vs. mixing competition is through comparison of volume of the *tidal prism* (tidal range times surface area) versus volume of freshwater input (e.g., Chapter 10). The result of the mixing vs. buoyancy effects ultimately determines the strength of water column stratification (Figure 1.3).

When tidal currents are relatively strong and river discharge is relatively weak, vertical stratification will be absent or relatively weak. This situation is analogous to tidal prisms overwhelming freshwater volumes from buoyancy input. The mismatch in volumes results in vertically mixed or vertically homogeneous semienclosed basins, such as coastal lagoons, as well as wide and shallow embayments. Expected

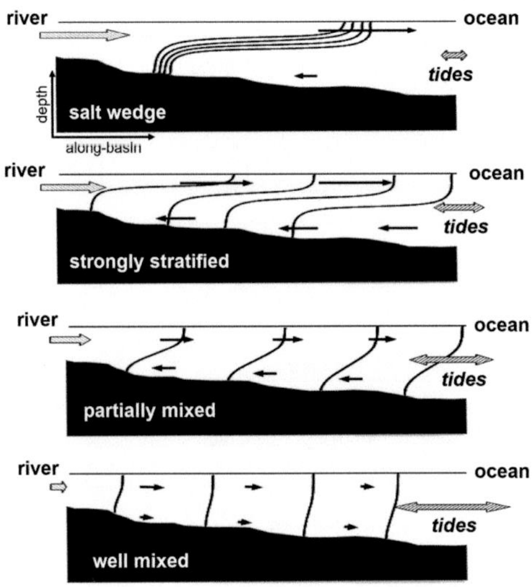

Figure 1.3 Classification based on water column stratification showing the four different types. Along-basin sections display mean density and flow fields. Freshwater volume decreases from top to bottom and tidal prism increases from top to bottom. Arrows are not scaled.

mean velocity profiles in vertically homogeneous basins are unidirectional with depth (Figure 1.3). Exchange flows will then be segregated laterally (laterally sheared), as studied in Chapters 7 and 8.

With river discharge increasing, vertical stratification should increase and result in weakly stratified or partially mixed basins. This occurs with moderate to strong tidal forcing and relatively weak to moderate river discharge. In this scenario, vertical stratification can vary greatly within one tidal cycle. Partially mixed conditions may be observed in lagoons or in estuaries with relatively low river discharge. Many temperate estuaries fall in this category and may become moderately stratified. The mean salinity profile displays a barely appreciable pycnocline or nearly continuous stratification throughout the water column. Mean flows tend to display the most vigorous volume exchange, compared to other stratification conditions (Figure 1.3). This vigorous exchange ensues from vertical mixing that maintains robust horizontal density gradients.

With further increased river discharge and comparatively decreased tidal forcing, the basin becomes strongly stratified. This situation can result from moderate to relatively large river discharge and relatively weak to moderate tidal forcing. Strong stratification (say surface-to-bottom salinity difference >5 g/kg) persists throughout the tidal cycle as in fjords and other deep (>20 m deep) estuaries. The mean salinity profile exhibits a pronounced halocline (pycnocline) with small (<2–3 g/kg) vertical variations in the profile above and below the halocline. Mean exchange flows are well developed but not as robust as in partially mixed basins because of comparatively reduced vertical mixing and decreased horizontal density gradients (Figure 1.3).

Salt-wedge basins develop with large river volume inputs compared to the tidal prism. There are examples of this type of basin throughout the world where tidal ranges are relatively small (microtidal basins). In most of these basins, salt wedges are markedly tidal. Flood tides wedge salty waters into a riverine basin. Wedge-type stratification appears in this tidal phase in which a layer of fresh water is separated by a bottom salty-water layer by a sharp pycnocline. Ensuing ebb tides are dominated by vertically homogeneous riverine waters. Therefore, salt wedges appear only in flood tides. In deep enough basins where the ebbing fresh waters remain from reaching the bottom, the wedge is semi-permanent and displays relatively short horizontal excursions.

1.3 Take-Home Message

As seen, any given semienclosed basin may change from vertically homogeneous to strongly stratified in one tidal cycle. These stratification changes can occur at various temporal scales and that is why it may sometimes be irrelevant to classify a basin with these guidelines unless we specify the gamut of variability and its temporal scales. A basin typically changes geomorphology over hundreds to thousands of years and its classification will likely remain the same in our lifetime. However, a basin may change water balances in monthly to seasonal scales and vertical stratification in even shorter periods. Because of this variability, we should be cautious when trying to classify any semienclosed basin. Other robust approaches to classifying these basins are presented in Chapter 12, which includes a dynamical approach. So, before understanding such an approach, we need to familiarize ourselves with the fundamentals of the dynamics of semienclosed basins. This acquaintance is formulated in the following Chapters (2–11).

Additional Sources

Geyer, W.R., and P. MacCready (2014) The estuarine circulation. *Ann. Rev. Fluid Mech.* 46: 175–197.

Guha, A., and G.A. Lawrence (2013) Estuary classification revisited. *J. Phys. Ocean.* 43 (8): 1566–1571.

Hansen, D.V., and M. Rattray Jr. (1966) New dimensions in estuary classification 1. *Limn. Ocean.* 11(3): 319–326.

Valle-Levinson, A. (2010) Definition and classification of estuaries. In *Contemporary Issues in Estuarine Physics*. Edited by A. Valle-Levinson, pp. 1–11. Cambridge: Cambridge University Press.

2

Conservation Equations

This chapter presents the basic quantitative tools used to study the hydrodynamics of semienclosed basins. In the most general sense, these equations apply to a wide variety of geophysical water motions. The chapter begins with the concept and equation of conservation of mass. Then it derives the quantitative description for conservation of salt and extends it to conservation of heat. It continues with a description of the Thermodynamic Equation of Seawater and the separate effects of temperature and salinity on water density. The chapter then presents conservation of momentum, which contains the hydrodynamic equations to study any problem in semienclosed basins (or the ocean, for that matter). The chapter concludes with an attempt to describe intuitively the frictional term in the momentum equations.

To benefit from this chapter, readers are asked to equip themselves with patience. The chapter attempts to present concepts in the most basic way so that readers without any calculus background can follow most materials. Consider that all equations shown here represent a shorthand description of concepts that can actually be imagined or that are familiar to readers. Equations are symbolic representations that explain ideas in an alternative language; try to regard them as your friends. At the same time, equations represent a mathematical problem that is solvable through various approaches. Therefore, they portray convenient ways to pose a problem that can be solved quantitatively. The study of hydrodynamics in a semienclosed basin is less complicated than the quantitative study of its biological, chemical, or sediment transport processes. This is because quantitative descriptions of hydrodynamics include less uncertainty than the quantitative depictions of the other aspects. So, if you think that the study of hydrodynamics is the most challenging, this chapter aims to show that it is the other way around.

2.1 Conservation of Mass

The most basic concept to study fluid motions is to assume that mass is conserved within a reference volume. To obtain an expression for *conservation of mass*, let us

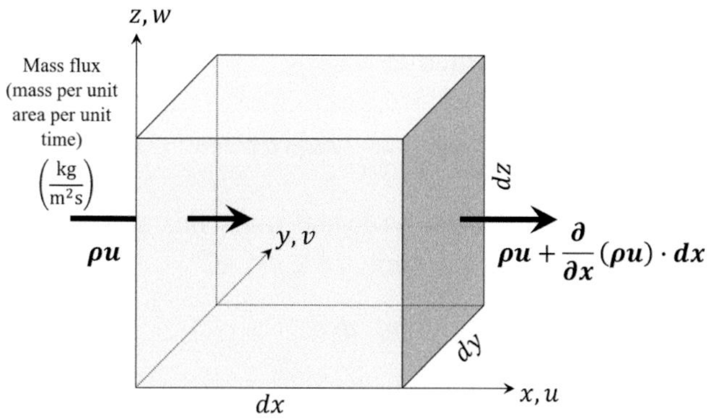

Figure 2.1 Volume element with reference frame and variables used for the derivation of conservation of mass.

first consider an infinitesimally small reference volume. To visualize such a volume, maybe you have to pretend that you take a pill that makes you the size of an atom. Furthermore, assume that the reference volume is your favorite room and that the room has a cubic shape (e.g., Figure 2.1). Next, assume that one of the floor edges of the room points in the east direction and arbitrarily assign this edge as the x axis (Figure 2.1). Then, represent an "L" shape with the index finger and the thumb of your right hand. Align your index finger with the x axis (the floor's edge). Your thumb will thus be aligned to the floor's edge that is perpendicular to the x axis. This room's edge (aligned with your thumb) should point toward the north and shall become the y axis (Figure 2.1). Finally, the corner of the room jutting upward at the intersection between the edges that denote the x and y axes will be called the z axis (see Figure 2.1). Therefore, you are inside this reference volume with dimensions dx, dy, dz. These dimensions are the length of the cubic room in each of the three directions established. In this context, because the volume element is a cube, $dx = dy = dz$.

Next, we are going to study fluxes of water into and out of this reference volume. We shall explore water motion with components u, v, and w in the x, y, and z directions, respectively (Figure 2.1). Further, we can safely say that mass is conserved in this volume, that is, the same water that enters the volume exits the volume. This is the essence of the conservation of mass concept. First, we can arbitrarily assign ρ as the density of water (mass per unit volume) and u the flow (m/s) that brings water mass into the volume element. The mass flux (mass per unit area per unit time) moving through the wall of the volume element (or room) with area $dydz$ is given by ρu [kg/(m^{2}s)]. It follows that the mass flux moving through the other wall of area $dydz$, parallel to the first wall, is the same mass flux that enters plus any changes to that flux with respect to the direction of motion (x), along the distance dx of the room. In symbols (or equation), the mass flux out of the volume element may be written as

$$\rho u + \frac{\partial}{\partial x}(\rho u) \cdot dx \tag{2.1}$$

(see Figure 2.1). We can therefore write the mass flux into the volume element throughout the area $dydz$ (mass per time, or kg/s) as

$$\rho u \cdot dydz \tag{2.2}$$

and the flux out (throughout the same area, i.e., equation 2.1 times $dydz$) as

$$\rho u \cdot dydz + \frac{\partial}{\partial x}(\rho u) \cdot dxdydz. \tag{2.3}$$

The net flux (kg/s) in the x direction is given then by equation 2.2 minus equation 2.3:

$$-\frac{\partial}{\partial x}(\rho u) \cdot dx\, dy\, dz. \tag{2.4}$$

By analogy, the net mass fluxes (kg/s) in the other two directions y and z are

$$-\frac{\partial}{\partial y}(\rho v) \cdot dx\, dy\, dz \text{ and} \tag{2.5}$$

$$-\frac{\partial}{\partial z}(\rho w) \cdot dx\, dy\, dz. \tag{2.6}$$

The net mass fluxes throughout the volume element actually describe the changes of mass M in the element, over time, that is, $\partial M/\partial t$, which can be written as

$$\frac{\partial M}{\partial t} = \frac{\partial}{\partial t}(\rho \cdot dx\, dy\, dz) = \left[-\frac{\partial}{\partial x}(\rho u) - \frac{\partial}{\partial y}(\rho v) - \frac{\partial}{\partial z}(\rho w)\right] \cdot dx\, dy\, dz \tag{2.7}$$

because mass equals density times volume ($M = \rho \cdot dx\, dy\, dz$). Then, if we write the net mass fluxes per unit volume ($dx\, dy\, dz$), equation 2.7 for conservation of mass becomes

$$\frac{\partial \rho}{\partial t} + \frac{\partial}{\partial x}(\rho u) + \frac{\partial}{\partial y}(\rho v) + \frac{\partial}{\partial z}(\rho w) = 0. \tag{2.8}$$

This is the most general expression for conservation of mass. Mass does not necessarily need to be conserved through the two walls, or sides, with the same orientation. This expression indicates that mass is conserved after we consider all sides of the volume element.

Expanding equation 2.8, that is, distributing the changes of a product, its equivalent is

$$\frac{\partial \rho}{\partial t} + u\frac{\partial \rho}{\partial x} + v\frac{\partial \rho}{\partial y} + w\frac{\partial \rho}{\partial z} + \rho\left[\frac{\partial u}{\partial x} + \frac{\partial v}{\partial y} + \frac{\partial w}{\partial z}\right] = 0. \tag{2.9}$$

The first four terms in equation 2.9 describe changes in density with respect to time (t) and space (x, y, z). Those four terms can be called the total changes (derivative) of density ($D\rho/Dt$), so that equation 2.9 becomes

$$\frac{1}{\rho}\frac{D\rho}{Dt} + \frac{\partial u}{\partial x} + \frac{\partial v}{\partial y} + \frac{\partial w}{\partial z} = 0. \tag{2.10}$$

This is another way of expressing conservation of mass. Equation 2.10 describes total changes of density (first term) and gradients in the velocity field. If we put numbers on the first and second terms in this equation, we find that $1/\rho\ D\rho/Dt$ is typically three orders of magnitude smaller than $\partial u/\partial x$ or the other two velocity gradients. Therefore, the changes of density in the ocean are negligible, compared to velocity gradients, unless the changes in density appear with gravity, as in baroclinic pressure gradients (Section 2.5). Such an assumption, that $1/\rho\ D\rho/Dt \approx 0$, is one form of expressing the *Boussinesq approximation.*

With the Boussinesq approximation, the expression for conservation of mass (equation 2.10) simply becomes

$$\frac{\partial u}{\partial x} + \frac{\partial v}{\partial y} + \frac{\partial w}{\partial z} = 0. \tag{2.11}$$

Equation 2.11 is another form of conservation of mass, known as *Continuity equation.* It also indicates that the flow is "non-divergent." Typically, this is the equation used in combination with *conservation of momentum* to solve any problem related to water mechanics, that is, hydrodynamics.

The concept of conservation of mass can be placed in the global context of a semienclosed basin by examining the fluxes of water that enter and leave the basin. From the most general perspective, water inputs (in m^3/s) would arise from precipitation P, river discharge R, thawing θ, groundwater discharge F_g, and volume inflow from the ocean F_o (Figure 2.2). Water outputs (or losses from the basin) would ensue from evaporation E, freezing $\emptyset$, volume outflow to the ocean F_b, and seepage through the sea bed F_a. Conservation of mass (or continuity), in this global context, would be (e.g., Figure 2.2)

$$R + P + F_o + \theta + F_g = E + F_b + \emptyset + F_a. \tag{2.12}$$

2.2 Conservation of Salt and Heat

For the concept of conservation of salt, we will place ourselves again in the microscopic volume element. We will consider fluxes of salt into and out of this

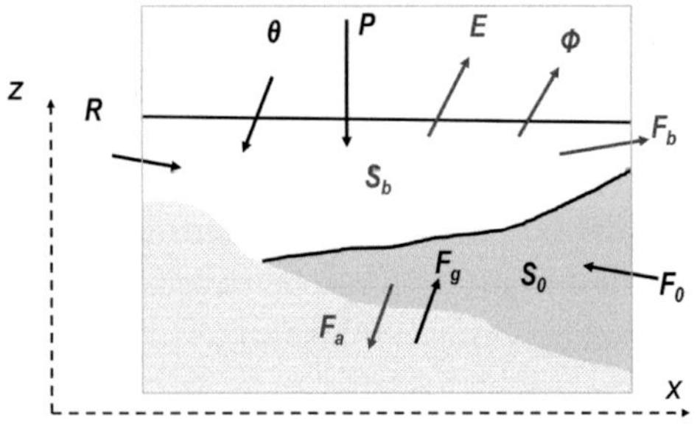

Figure 2.2 Longitudinal section of a generic basin illustrating the global concept of conservation of mass, showing possible volume inputs (left-hand side of equation 2.12) and outputs (right-hand-side of equation 2.12). Oceanic water of salinity S_0 is shaded compared to basin's water of salinity S_b.

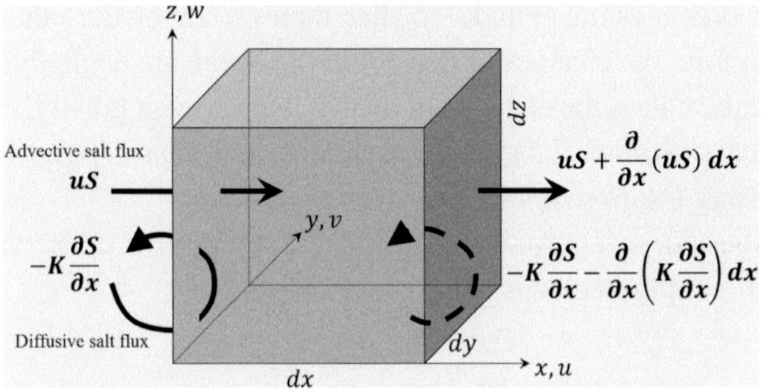

Figure 2.3 Volume element with reference frame and variables used for the derivation of conservation of salt.

volume element. In this case (Figure 2.3), we can assume that the volume element has higher salinity S or salt concentration (g/kg) than the surroundings. Fluxes of salt will now have two contributions: (i) an advective salt flux related to the transport of salt by the flow (Figure 2.3) and (ii) a diffusive salt flux linked to the salinity contrast between the volume element and its surroundings. The advective salt flux in the x direction is uS and effectively describes the transport of salt per unit area per unit time (salt flux). We will assume that the diffusive flux is turbulent without going into the details of turbulence. Even without a flow u, salinity differences will tend to smooth out through diffusive processes. This diffusive flux has the same units as the advective flux and is proportional to the salinity difference, which can be represented by $\partial S/\partial x$ in the x direction. The diffusive flux has a negative sign because it develops against the gradient $\partial S/\partial x$ and has a "constant" of proportionality K. The constant represents salt diffusivity and has units of velocity times length (m^2/s). In a way, this constant may be thought of as the speed of the diffusivity eddies that promote salt diffusive fluxes multiplied by the size of those eddies. In other words, the diffusivity "constant" (in reality K varies in time and space) represents the efficacy of the diffusive processes.

Following the same arguments as for conservation of mass, the salt flux into the volume element in the x direction is given by

$$\text{flux in} = \underbrace{uS}_{advective} \quad \underbrace{-K\frac{\partial S}{\partial x}}_{diffusive} . \tag{2.13}$$

Similarly, the flux out of the volume element (Figure 2.3) is given by the same flux into the volume plus any of its changes with respect to the x direction along the distance dx of the volume element:

$$\text{flux out} = \underbrace{uS + \frac{\partial}{\partial x}(uS)dx}_{advective} \underbrace{-K\frac{\partial S}{\partial x} - \frac{\partial}{\partial x}\left(K\frac{\partial S}{\partial x}\right)dx}_{diffusive} . \tag{2.14}$$

The net salt flux in the x direction, over the entire surface $dydz$, is given by equation 2.13 minus equation 2.14 times $dydz$:

$$\text{net flux in } x = \underbrace{-\frac{\partial}{\partial x}(uS)dx\,dy\,dz}_{advective} + \underbrace{\frac{\partial}{\partial x}\left(K\frac{\partial S}{\partial x}\right)dx\,dy\,dz}_{diffusive} . \tag{2.15}$$

Through the same reasoning, the net salt fluxes in the y and z direction are therefore given by

$$-\frac{\partial}{\partial y}(vS)dx\,dy\,dz + \frac{\partial}{\partial y}\left(K\frac{\partial S}{\partial y}\right)dx\,dy\,dz \tag{2.16}$$

and

$$-\frac{\partial}{\partial z}(wS)dx\,dy\,dz + \frac{\partial}{\partial z}\left(K_z\frac{\partial S}{\partial z}\right)dx\,dy\,dz. \tag{2.17}$$

Thus, the net salt flux per unit volume $(dx\,dy\,dz)$, or $\partial S/\partial t$, may be expressed as

$$\frac{\partial S}{\partial t} = -\frac{\partial}{\partial x}(uS) - \frac{\partial}{\partial y}(vS) - \frac{\partial}{\partial z}(wS) + \frac{\partial}{\partial x}\left(K\frac{\partial S}{\partial x}\right) + \frac{\partial}{\partial y}\left(K\frac{\partial S}{\partial y}\right) + \frac{\partial}{\partial z}\left(K_z\frac{\partial S}{\partial z}\right). \tag{2.18}$$

The first three terms on the right-hand side of equation 2.18 are equivalent to

$$\frac{\partial}{\partial x}(uS)+\frac{\partial}{\partial y}(vS)+\frac{\partial}{\partial z}(wS)=u\frac{\partial S}{\partial x}+v\frac{\partial S}{\partial y}+w\frac{\partial S}{\partial z}+S\left[\frac{\partial u}{\partial x}+\frac{\partial v}{\partial y}+\frac{\partial w}{\partial z}\right]. \quad (2.19)$$

The term in brackets on the right-hand side of equation 2.19 denotes Continuity (equation 2.11) and equals zero. Therefore, conservation of salt is expressed as (from equations 2.18 and 2.19)

$$\frac{\partial S}{\partial t}+u\frac{\partial S}{\partial x}+v\frac{\partial S}{\partial y}+w\frac{\partial S}{\partial z}=\frac{\partial}{\partial x}\left(K\frac{\partial S}{\partial x}\right)+\frac{\partial}{\partial y}\left(K\frac{\partial S}{\partial y}\right)+\frac{\partial}{\partial z}\left(K_z\frac{\partial S}{\partial z}\right). \quad (2.20)$$

This equation says that the total changes of salinity in time (the first four terms together), that is, local plus advective changes in salinity, are equal to the diffusive changes. If we assume steady state (negligible local changes) and constant diffusivities, equation 2.20 becomes

$$u\frac{\partial S}{\partial x}+v\frac{\partial S}{\partial y}+w\frac{\partial S}{\partial z}=K\frac{\partial^2 S}{\partial x^2}+K\frac{\partial^2 S}{\partial y^2}+K_z\frac{\partial^2 S}{\partial z^2}. \quad (2.21)$$

This is the *Advection–Diffusion* equation for salt, explaining salt fluxes (g/kg/s) integrated over a cross-sectional area. Recall that a flux indicates changes of a property over time and per unit area. In equations 2.17–2.21 (except for 2.19), horizontal diffusivity of salt K is distinguished from vertical diffusivity K_z, the latter being typically two-to-three orders of magnitude smaller than the former. Also, without going into the details of turbulence, it is assumed that the diffusivity coefficients are turbulent.

Further, assuming volume-integrated fluxes in one direction, the statement of conservation of salt simply becomes (see Figure 2.2)

$$F_o S_o = F_b S_b \quad (2.22)$$

The volume-integrated conservation equations (2.12 and 2.22) are simple but powerful. With the net discharge R into a basin, neglecting other gains and losses, and the values of salinity related to inflows and outflows (S_o, S_b), we can determine the values of F_o and F_b. The combined equations 2.12 and 2.22 are also known as *Basin Equations* or *Knudsen's Equations*. They describe a system of two algebraic simultaneous equations that can be solved for F_o and F_b with the information just mentioned. The concept has been applied successfully to semienclosed basins, obtaining at least the correct order of magnitude of F_o and F_b.

For *conservation of heat*, we may follow the same derivation as for *conservation of salt*, considering an advective and a diffusive flux of heat. Recall that heat (in Joules – J – or kg m^2/s^2) is the energy related to the temperature of a substance.

It equals mass of the substance m times specific heat capacity C_a (heat capacity per unit mass, which is $\approx$ 4,000 J/[kg °C] for sea water) times its temperature T, that is, $m \cdot C_a \cdot T$. Thus, in the derivation outlined in equations 2.12–2.20, instead of salinity S we can use heat $m \cdot C_a \cdot T$ or heat per unit volume $\rho \cdot C_a \cdot T$. Because mass and specific heat capacity appear in every term, the conservation of heat expression is

$$\frac{\partial T}{\partial t} + u\frac{\partial T}{\partial x} + v\frac{\partial T}{\partial y} + w\frac{\partial T}{\partial z} = \frac{\partial}{\partial x}\left(K\frac{\partial T}{\partial x}\right) + \frac{\partial}{\partial y}\left(K\frac{\partial T}{\partial y}\right) + \frac{\partial}{\partial z}\left(K_z\frac{\partial T}{\partial z}\right) \tag{2.23}$$

Heat exchanges with the atmosphere represent the diffusive flux $\kappa_z \partial T/\partial z$ at the air–water interface. In other words, heat fluxes at the surface enter a problem as a boundary condition representing the vertical diffusive flux of heat. It is customary to parametrize these surface heat fluxes through widely used equations for short- and long-wave radiation, latent and sensible fluxes.

Equations 2.20 and 2.23 show that conservation of heat and salt have the same form. A key distinction is in the diffusivity coefficients. Heat diffuses faster than salt and, like for salt, κ is typically two-to-three orders of magnitude greater than κ_z. In most estuaries, salinity will overwhelm heat (temperature) in affecting the density field. There are many semienclosed basins, however, where temperature may also play a role or even dominate (Chapter 11). The relative influence of salinity and temperature on density is explored with the Thermodynamic Equation of Seawater.

2.3 Thermodynamic Equation of Seawater

This is an equation of state that relates temperature T, salinity S, and pressure (depth) p to water density ρ. The most generic form of this expression is

$$\rho = \rho(S, T, p) \tag{2.24}$$

A polynomial relationship that satisfies equation 2.24 is given in the "International thermodynamic equation of seawater – 2010," advanced by the Intergovernmental Oceanographic Commission. This equation has evolved over nearly four decades of measuring tetrads of S, T, p, and ρ in the laboratory to arrive at a thermodynamically consistent relationship. The equation is known as *TEOS-10* and its lengthy polynomial has 48 coefficients and different combinations of S, T, and p. In coastal bodies of water, at depths of <500 m, the effects of pressure on water density are negligible. So, for practical purposes, p equals 0 in estimates of ρ.

More information on this thermodynamic equation, including useful routines in different programming languages, may be found by simply searching "teos-10." Among the most relevant concepts to extract from TEOS-10 is that salinity values should be reported as *Absolute Salinity* S_A, with units of g/kg. Values of S_A in coastal waters are the same as those of Reference Salinity S_R (in g/kg), down to a precision of 0.1 g/kg. In turn, S_R is related to the Practical Salinity S_P (unitless) values that are measured with a Conductivity-Temperature-Depth (CTD) recorder via

$$S_R = \frac{35.16504}{35} S_P = 1.00471\, S_P \tag{2.25}$$

In semienclosed coastal bodies of water, we are mostly concerned about gradients in salinity, as they are possible dynamic agents in the baroclinic pressure gradient. This dynamic influence of salinity gradients is described in Section 2.5. So, for dynamical and practical intents and purposes, it should be fine to assume

$$S_A \approx S_R \approx S_P \tag{2.26}$$

that is, report the salinity measured with a CTD as S_A in g/kg. In the rest of this text, we simply refer to salinity as S with implicit units of g/kg, remembering that derived salinity from conductivity is actually S_P.

2.4 Relative Influence of Salinity and Temperature on Density

Drawing water density values ρ as a function of T and S (for $p = 0$) with the Thermodynamic Equation of Seawater (TEOS-10, Figure 2.4) identifies the qualitative influence of each. The reader should attempt to use the Matlab script provided at the end of the chapter to get a better feel for TEOS-10. Inspection of Figure 2.4 (a T–S diagram) indicates that changes of ρ are faster in the horizontal direction (with S) than in the vertical direction (with T). It also shows that ρ changes with T (for fixed S) are smaller at low temperatures (<15 °C) than at high temperatures (>20 °C). The figure thus reveals that salinity is relatively more influential than temperature in determining water density.

The quantitative influence of T or S on ρ is determined with the coefficient of thermal expansion α for temperature, and with the coefficient of saline or haline contraction β for salinity. Thermal expansion α explains how much the water density changes with varying water temperature:

$$\alpha = -\frac{1}{\rho}\left(\frac{\partial \rho}{\partial T}\right) \tag{2.27}$$

Equation 2.27 represents what we just wrote in words: how much density changes ($\partial \rho$) relative to temperature changes (∂T). The minus sign in equation 2.27 says that a positive change in temperature will cause a density decrease (for $T > 4$ °C). Using TEOS-10, we can plot values of α as a function of S and T (Figure 2.5a, see

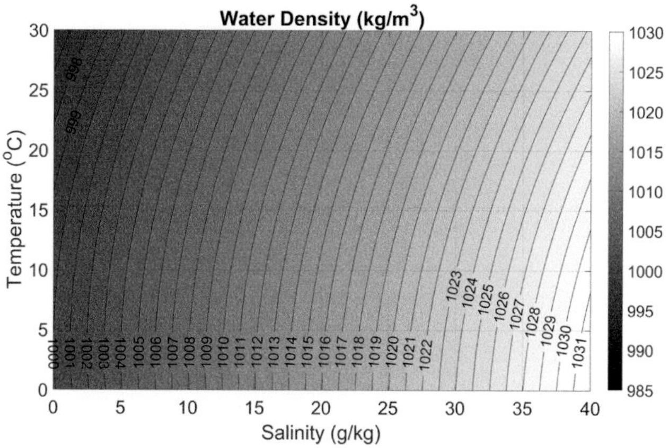

Figure 2.4 T–S diagram showing values of water density as a function of temperature and salinity at the surface.

also Matlab script provided at the end of this chapter). As an illustration of the meaning of α, negative values (lower-left corner of Figure 2.5a) denote density increasing with an increase in temperature, and vice versa. Then, at high values of S and T, at the upper-right corner of Figure 2.5a, α values are near 3×10^{-4} °C^{-1}. This α value indicates that an increase in water temperature of 1 °C, within those T–S combinations, will cause a decrease in water density of 0.3 kg/m^3. Maximum effects of temperature on density occur in that parametric region. In general, we say that α is typically 2×10^{-4} °C^{-1} (a rough, representative value of Figure 2.5a), which would describe a density change of 0.2 kg/m^3 in the opposite direction (with the opposite sign) as the temperature changes by 1 °C.

Similarly, the coefficient β is determined quantitatively through

$$\beta = \frac{1}{\rho}\left(\frac{\partial \rho}{\partial S}\right) \tag{2.28}$$

indicating how much ρ changes with a change in salinity ∂S. Again, we use TEOS-10 routines to determine β (Figure 2.5b). The range of variation for β is much smaller than for α as it can acquire values between 7.2 and 8.1×10^{-4} kg/g. The lowest values of β occur for the highest temperatures and salinities in the parameter space on the figure. In general, we can say that β is roughly four times greater than α, meaning that salinity is that much more effective than temperature in influencing density.

A detailed inspection of the ratio $|\beta/\alpha|$ (where the vertical bars || denote absolute value, Figure 2.5c) shows that salinity is at least three times more important than temperature. This is restricted to high water temperatures (upper-right corner). In most of the parameter space, however, salinity is >4 times more

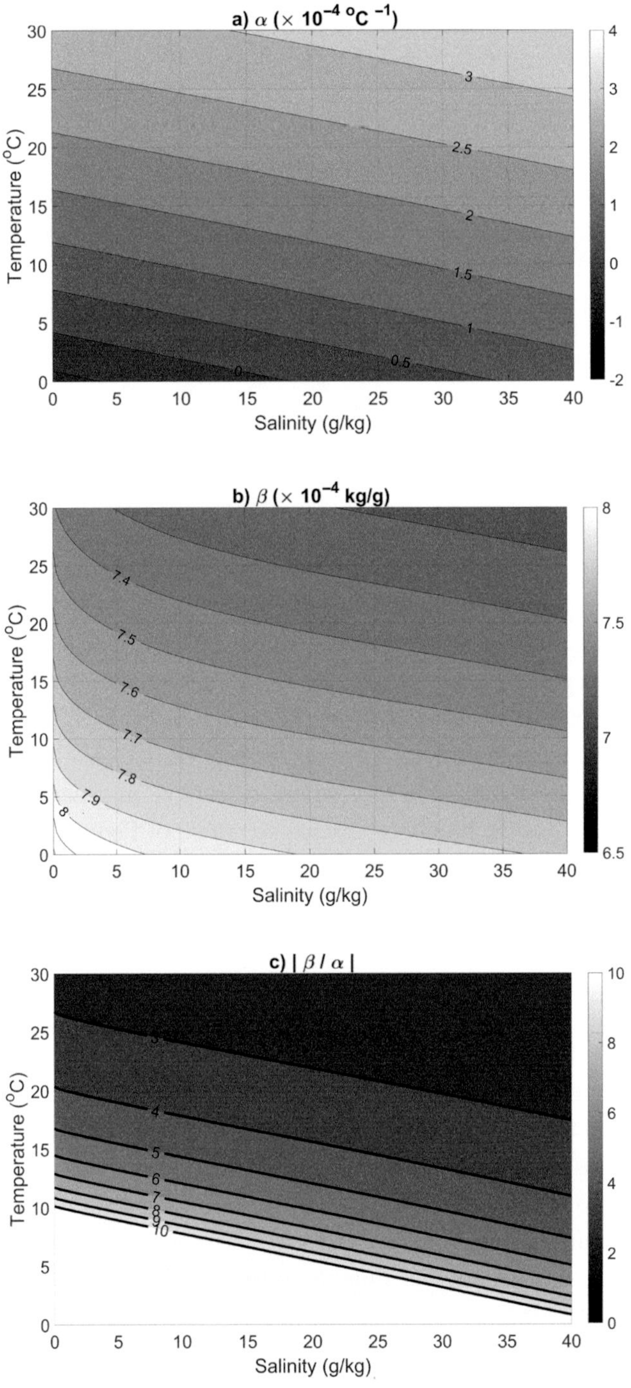

Figure 2.5 Coefficients of (a) thermal expansion $\alpha\left(\times 10^{-4}\right)$, (b) haline contraction $\beta\left(\times 10^{-4}\right)$, and (c) the absolute value of the ratio β/α. All coefficients are drawn as a function of temperature and salinity values and for $p = 0$.

influential than temperature. Salinity dominates by >10 times at low temperatures and salinities (lower-left corner).

It is evident that temperature's role on density may become prominent at relatively high water temperatures, as in tropical and subtropical semienclosed basins. There, gradients in temperature may play a dynamical role. In temperate and high-latitude estuaries, where salinity dominates, the equation of state that is normally used disregards temperature influence and is approximated as

$$\rho = \rho_f(1 + \beta S) \approx \rho_f + 0.75\, S \tag{2.29}$$

where ρ_f is freshwater density (1,000 kg/m^3). In other words, $\rho_f \beta S$ is the increase in water density, relative to fresh water, caused by salinity. Keep in mind that equation 2.29 is valid for estuaries whose temperature–salinity pairs fall away from the upper-right corner of Figure 2.5c.

Up to now, we have collected four equations that permit the study of water motions in a semienclosed basin. These equations are *conservation of mass* (2.11), *conservation of salt* (2.20), *conservation of heat* (2.23), and *equation of state* (2.24). The three conservation equations are interconnected to the flow field u, v, w, and the equation of state establishes the interrelationships between density, temperature, salinity, and pressure. Next section explores the dynamic representation of the flow field as well as the dynamic influences of temperature and salinity gradients, through *conservation of momentum.*

2.5 Conservation of Momentum

The expression for conservation of momentum originates from a fundamental law in physics, the mother of all laws in the study of semienclosed basin dynamics, namely Newton's second law:

$$m\,\vec{a} = \sum \vec{F} \tag{2.30}$$

where the arrow above the letters indicates three-dimensional accelerations and forces, which are vectors with three components. Newton's second law can alternatively be written in terms of accelerations (units m/s^2 or force/unit mass or Newtons/kg):

$$\vec{a} = \sum \frac{\vec{F}}{m}. \tag{2.31}$$

We digress briefly to explain why we refer to these equations (three equations because there are three components in the acceleration and forces) as conservation of momentum. We know that *momentum* is mass times velocity ($m\,\vec{v}$). If we explore momentum fluxes (just like we studied mass fluxes in Figure 2.1) per unit mass through a volume element, we end up with changes in velocity per unit time or accelerations. Therefore, equation 2.31 is an equation of conservation of momentum flux (throughout a cross-sectional unit area) per unit mass, or conservation of momentum per unit time and mass (acceleration), or simply conservation of momentum.

Returning to equation 2.31, it tells us that accelerations on any fluid portion (left-hand side) are produced by the sum of all forces per unit mass (right-hand side) acting on that fluid. To present a tractable expression for 2.31, first we expand the expression that describes acceleration, and then we represent the forces per unit mass that cause acceleration.

The left-hand side of equation 2.31 has three components, each containing a local acceleration and advective accelerations. Therefore, from a single symbol, the left-hand side of 2.31 expands to:

$$\vec{a} = \frac{d\vec{v}}{dt} = \begin{bmatrix} \frac{\partial u}{\partial t} + u\frac{\partial u}{\partial x} + v\frac{\partial u}{\partial y} + w\frac{\partial u}{\partial z} \\ \frac{\partial v}{\partial t} + u\frac{\partial v}{\partial x} + v\frac{\partial v}{\partial y} + w\frac{\partial v}{\partial z} \\ \frac{\partial w}{\partial t} + u\frac{\partial w}{\partial x} + v\frac{\partial w}{\partial y} + w\frac{\partial w}{\partial z} \end{bmatrix} \tag{2.32}$$

But that is not all for the left-hand side. If we study accelerations from afar (from outer space) by examining a reference frame that rotates, as an observer from space of a fluid motion on Earth, we need to include *Coriolis accelerations*. Derivation of these accelerations falls outside the scope of this text. Suffice it to say that the three components of *Coriolis* accelerations are

$$[-fv,\ fu,\ 0] \tag{2.33}$$

The vertical component is not exactly zero but is negligible because it is several orders of magnitude smaller than *gravity*, which is the dominant acceleration in the vertical. In equation 2.33, f is the *Coriolis parameter* and equals

$$f = 2\Omega \sin \Lambda \tag{2.34}$$

where Ω is the rate of Earth's rotation ($2\pi/24$ hours or $2\pi/86400$ seconds), and Λ is the latitude of the study area (negative in the Southern Hemisphere). Thus, f ranges from zero at the Equator to $1.45 \times 10^{-4}\ \mathrm{s}^{-1}$ (maximum) at the poles. In geophysical fluid dynamics, changes of f with latitude Λ should be accounted for. However, in most semienclosed basins we can assume f is constant for a particular Λ.

The right-hand side of equation 2.31 includes three main forces per unit mass that cause accelerations (positive or negative): *pressure gradients*, *gravitational force*, and *friction*. We can derive the terms describing accelerations from pressure gradient in the same way as we derived conservation of mass, instead looking at pressure on a volume element. This yields, in the x direction, the term

$$-\frac{1}{\rho}\frac{\partial p}{\partial x} \tag{2.35}$$

We can also take the definition of hydrostatic pressure that the pressure at any depth z (positive upward), measured from the surface η, is given by

$$p = -g\int_{z}^{\eta} \rho \, dz \tag{2.36}$$

where g is the gravitational field force per unit mass or gravity's acceleration. Inserting equation 2.36 in equation 2.35 provides the pressure gradient forces per unit mass in the x direction:

$$-\frac{1}{\rho}\frac{\partial p}{\partial x} = -g\frac{\partial \eta}{\partial x} - \int_{-H}^{z} \frac{g}{\rho}\frac{\partial \rho}{\partial x} dz \tag{2.37}$$

The first term on the right-hand side is the acceleration caused by the sea-surface slope $\partial\eta/\partial x$ and the second term is related to horizontal density gradients $\partial\rho/\partial x$. The second term is known as the baroclinic pressure gradient and can be influenced mainly by salinity gradients in estuaries (Section 2.4), but also by temperature gradients in low-discharge basins (Chapter 11).

In the y direction, accelerations caused by the pressure gradient have the same form as equation 2.37 but with y instead of x. Even if the water density is constant in the vertical (vertically uniform), the baroclinic pressure gradient increases with depth and can play a dynamic role if there are non-negligible horizontal density gradients. The vertical component of the pressure gradient is simply $-\frac{1}{\rho}\frac{\partial p}{\partial z}$.

Gravitational forces provide tides in the horizontal momentum and gravity acceleration in the vertical. Tides enter a problem as a boundary condition and are treated in detail in Chapters 3–5. Acceleration due to gravity has three components:

$$\vec{g} = [0, \quad 0, \quad -g] \tag{2.38}$$

where g is assumed constant, for practical purposes, and equals 9.8 m/s^2 (or 9.8 Newtons/kg).

Friction may be thought of as *diffusivity of momentum* and so it has a form that is similar to the diffusive terms in equations 2.20 and 2.23. Recall that such diffusivity is assumed turbulent, as we will skip the details of turbulence concepts. In the x direction, the diffusivity of momentum is represented by:

$$\frac{\partial}{\partial x}\left[A_h\frac{\partial u}{\partial x}\right]+\frac{\partial}{\partial y}\left[A_h\frac{\partial u}{\partial y}\right]+\frac{\partial}{\partial z}\left[A_z\frac{\partial u}{\partial z}\right] \tag{2.39}$$

Equation 2.39 is associated with the x momentum balance. For the y momentum balance, the diffusivity of momentum has the same form but with the v component instead of u. Both the horizontal eddy diffusivity A_h (m^2/s) and the vertical eddy diffusivity A_z (m^2/s) are assumed invariant in the x and y momentum balances. In the vertical momentum balance, momentum diffusivity terms tend to be much smaller than gravity and are usually neglected.

As seen in the next section, the terms in brackets in equation 2.39 indicate stresses in three directions as also denoted by $\vec{\tau_x}=\left(\tau_{xx},\tau_{xy},\tau_{xz}\right)$. Each term in brackets describes a gradient in the u velocity that represents u momentum transfer in the x, y, and z directions. All terms in equation 2.39 effectively denote frictional effects (spatial gradients of momentum transfers) via *stress divergence* ($\nabla\cdot\vec{\tau_x}$). Therefore, frictional terms are also referred to as stress divergence terms.

Consequently, in the study of semienclosed basins, the three momentum conservation equations look like (the expansion of equation 2.31):

$$\frac{\partial u}{\partial t}+u\frac{\partial u}{\partial x}+v\frac{\partial u}{\partial y}+w\frac{\partial u}{\partial z}-fv=-g\frac{\partial\eta}{\partial x}-\frac{g}{\rho}\int_{-H}^{z}\frac{\partial\rho}{\partial x}dz+\frac{\partial}{\partial x}\left[A_h\frac{\partial u}{\partial x}\right]+\frac{\partial}{\partial y}\left[A_h\frac{\partial u}{\partial y}\right]+\frac{\partial}{\partial z}\left[A_z\frac{\partial u}{\partial z}\right]$$

$$\frac{\partial v}{\partial t}+u\frac{\partial v}{\partial x}+v\frac{\partial v}{\partial y}+w\frac{\partial v}{\partial z}+fu=-g\frac{\partial\eta}{\partial y}-\frac{g}{\rho}\int_{-H}^{z}\frac{\partial\rho}{\partial y}dz+\frac{\partial}{\partial x}\left[A_h\frac{\partial v}{\partial x}\right]+\frac{\partial}{\partial y}\left[A_h\frac{\partial v}{\partial y}\right]+\frac{\partial}{\partial z}\left[A_z\frac{\partial v}{\partial z}\right]$$

$$0=\frac{1}{\rho}\frac{\partial P}{\partial z}+g \tag{2.40}$$

To review, the top two equations describe horizontal momentum (really accelerations) in the horizontal plane (x, y). The terms on the left-hand side of the top two equations represent accelerations (local, advective, and *Coriolis*) on a rotating frame. The first two terms on the right-hand side indicate accelerations caused by pressure gradients (from water-level slopes and horizontal density gradients). The rest of the terms illustrate stress divergence or frictional effects.

The lower-most equation in 2.40 is the momentum balance in the vertical direction, which becomes the *hydrostatic approximation*. This means that the dynamics are hydrostatic (negligible vertical flow) because local and advective accelerations in the vertical are negligible, as well as frictional effects, compared to gravity (in most cases). In instances where vertical flows are sizable (say > 0.1 m/s) the flow will likely be non-hydrostatic.

Equations 2.40 illustrate one form of the Reynolds-Averaged Navier–Stokes equations. These are the equations, together with equations 2.10, 2.20, 2.23, and 2.24, that we want to solve in any hydrodynamics study in semienclosed coastal basins and are the equations solved by numerical models. The Navier–Stokes equations, themselves, involve molecular viscosities ($\nu \approx 1 \times 10^{-6} m/s^2$ for water) instead of eddy viscosities A_h and A_z. Equations 2.40 are obtained by assuming turbulent flows in the Navier–Stokes equations. These turbulent flows are composed of a temporal mean (say 10-minute average) and a fluctuating deviation from that mean. Arriving at equations 2.40 from the Navier–Stokes equations is beyond the scope here but it is relatively straightforward. In the next section, we explore alternative representations for the frictional terms, in terms of the stresses (terms in brackets in equations 2.40).

2.6 Frictional or Stress Divergence Terms

To use equations 2.40 in the study of coastal waters, our greatest challenge is to find an approximation to the stress terms (those in brackets). The use of such approximation frequently represents the source of the largest uncertainty in the attempt to solve equations 2.40. Greater challenges arise in the attempt to study dissolved or suspended concentrations, together with equations 2.40. There are greater uncertainties when we try to approximate sources or sinks that should be included in conservation equations (such as 2.20 and 2.23) for dissolved or suspended matter concentrations. Thus, it is evident that the study of hydrodynamics poses less challenges than the study of any dissolved or suspended matter in the ocean.

Returning to equations 2.40, we usually neglect the stresses that redistribute momentum in the horizontal direction, that is, those that have the eddy viscosity represented by A_h. This assumption can be justified as follows. Values of A_h typically range between 1 and 100 m^2/s and A_z is customarily found between 1×10^{-4} and 1×10^{-2} m^2/s. Horizontal friction may be scaled as A_h/L_x^2, where L_x is a horizontal scale of motion, while vertical friction may be scaled as A_z/H^2, where H is a vertical scale of motion (e.g., depth). The ratio $\left[A_h/L_x^2\right]/\left[A_z/H^2\right]$ is typically <0.01, which makes horizontal friction negligible relative to vertical friction. In very energetic flows (>2 m/s) horizontal eddies will make horizontal friction dynamically relevant.

Our main concern then becomes how to represent the last of the frictional terms, those symbolizing a vertical transfer of horizontal momentum. We can think of these terms as eddies that redistribute momentum vertically because of vertical shears (Figure 2.6). This can happen at the air–water interface, at the water–bottom interface, and in the interior of the water column.

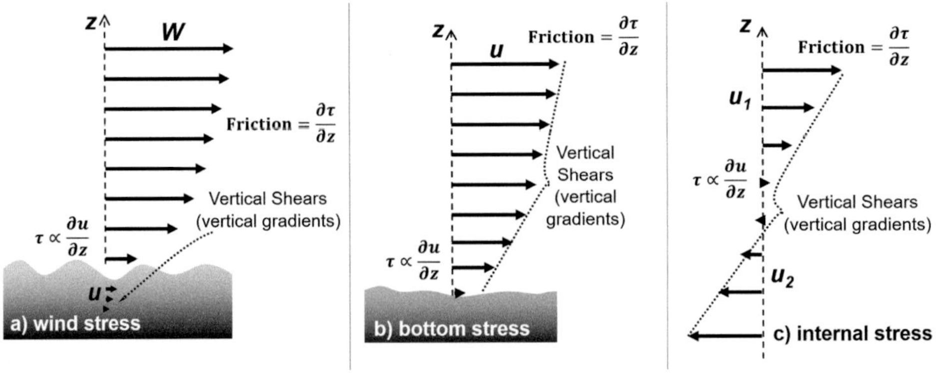

Figure 2.6 Schematic of vertical exchange of horizontal momentum from (a) wind stress at the surface, (b) bottom stress, and (c) internal stresses. Stresses are proportional to the vertical shear in horizontal velocity. Frictional effects are given by vertical gradients of these shears.

As wind transmits horizontal momentum in the vertical direction to the water (Figure 2.6a), the upper layers of the water column in turn transfer momentum in the vertical direction. Horizontal motions thus induced display vertical shears. Furthermore, the vertical gradients associated with the vertical shears represent frictional effects. The wind loses momentum to the water, which begins moving as it transfers its momentum in the vertical direction. In the x direction at the surface ($z = 0$, Figure 2.6a), the wind stress τ_s is traditionally parametrized as

$$A_z \frac{\partial u}{\partial z} = \frac{\tau_s}{\rho} = \frac{\rho_a C_d W_x |\vec{W}|}{\rho} \tag{2.41}$$

where ρ_a is the density of air (approximately 1.2 kg/m^3), ρ is water density, C_d is a dimensionless air–water drag coefficient, W_x is the east component of the wind velocity, and $|\vec{W}|$ is the wind speed. There are different ways of defining C_d (with typical order 10^{-3}) as a function of wind speed. This is an area of research with constant reexamination.

Similar to the air–water interface, at the water–bottom interface (Figure 2.6b) the vertical shears caused by the solid (or semi-solid) bottom generate a stress. The vertical gradients of such stress denote bottom frictional effects. Customary parametrizations of bottom stress (at $z = -H$) are given by

$$A_z \frac{\partial u}{\partial z} = \frac{\tau_b}{\rho} = \frac{\rho C_b u |\vec{V}|}{\rho} = C_b u |\vec{V}| \tag{2.42}$$

where ρ is water density, C_b is a nondimensional water–bottom drag coefficient, u is the east component of the current velocity, and $|\vec{V}|$ is the current speed. Velocity values are typically those at 1 m above the bottom. There are different values of C_b as a function of bottom roughness, but there is a canonical value of 0.0025. In "rough" bottoms such as coral, seagrass, or macroalgae, values of C_b can be one or more orders of magnitude larger.

In the interior of the water column, vertical shears in the horizontal flow will drive a vertical exchange of horizontal momentum (Figure 2.6c) that may be counteracted by stratification in the water column. Parametrization of internal stresses is probably the biggest challenge out of the three stresses illustrated in Figure 2.6. Many parametrizations use energetics arguments to approximate stress values. These parametrizations anchor on the competition between *stratifying* energy (per unit mass and area) from buoyancy versus *destratifying* energy (per unit mass and area) from vertical shears (shear production). Internal stress becomes a function of the nondimensional *Richardson Number* (*Ri*, buoyancy/shear production):

$$A_z \frac{\partial u}{\partial z} \approx f[Ri]; \; Ri = \frac{-\frac{g \partial \rho}{\rho \partial z}}{\left[\frac{\partial u}{\partial z}\right]^2 + \left[\frac{\partial v}{\partial z}\right]^2} \tag{2.43}$$

As instrument technology advances, we are becoming able to measure these stresses directly (Reynolds stresses). Measurements of the stress divergence terms in the momentum equations (2.40) are becoming, or soon will be, common in studies of semienclosed basins. However, Reynolds stresses in numerical models are coded directly only in domains of order 100 m. In the near future, these stresses will be represented in more realistic domains as computer capability advances.

2.7 Take-Home Message

The basic quantitative tools to study the hydrodynamics, and any suspended and dissolved matter in water, of semienclosed basins are the conservation equations. *Conservation of momentum* and *conservation of mass* are used for water motion, while *conservation of salt* and *conservation of heat* address their distribution in space and time. An equation of state, the Thermodynamic Equation of Seawater, relates temperature (heat content in the water column), salinity, and pressure to water density, whose gradients play a dynamic role in the conservation of momentum. The *conservation of suspended or dissolved matter* has an advective contribution that depends on water motion, a diffusive contribution, and a source/sink contribution. The source/sink contribution represents the greatest uncertainty in water-related studies.

Additional Sources

IOC, SCOR and IAPSO (2010) The international thermodynamic equation of seawater – 2010: Calculation and use of thermodynamic properties. Intergovernmental Oceanographic Commission, Manuals and Guides No. 56, UNESCO (English), 196 pp.

Pond, S., and G.L. Pickard (1983) *Introductory Dynamical Oceanography*. 2nd ed. Amsterdam: Elsevier.

3

Tides in Semienclosed Basins

This chapter covers basic concepts on the forces that drive tides. It presents a description of the main harmonics or constituents related to the driving forces, followed by a presentation of the concept of *spring* and *neap* tides, as well as their analogous *tropic* and *equatorial* tides. It then introduces the fundamental physics for a *frictionless tidal wave*, while depicting *progressive* and *standing* waves. It continues with the conditions for tidal *resonance*. The chapter then includes the effects of *bottom friction* on tidal currents and covers the effects of *convergent coastlines* by presenting the concepts of *hypersynchronic*, *hyposynchronic*, and *synchronic* basins. It follows with the effects of *Earth's rotation* on frictionless tides and the generation of *amphidromic* points. The chapter goes further with an exploration of the effects of lateral bathymetry on tidal flows.

3.1 Basic Concepts on Driving Forces

Tides are the most influential and predictable of the driving forces in most semienclosed basins. The first step in studying any semienclosed basin, regardless of study discipline, should be the understanding of tidal amplitudes and currents throughout the basin. This is because tides will determine fundamental variability in space and time throughout. We deal here with the basic concepts, leaving details to other texts that concentrate on this topic.

Tides essentially result from the balance (actually, slight imbalance) between gravitational attraction and centripetal forcing. Gravitational attraction will follow Newton's Law of Universal Gravitation that argues that two celestial bodies are attracted to each other with a force that is proportional to the product of their masses and inversely proportional to the square of the distance between them. The greatest gravitational pulls on Earth are exerted by the Moon and by the Sun. On the other hand, a centripetal force prevents collision between planets or moons induced by their gravitational attraction. The centripetal force arises when considering the Earth–Moon system, which revolves in space around a common center of mass (or center of gravity).

The tide-generating force, which is different from the gravitational force, is the difference between gravitational and centripetal forces. This gives rise to the concept of equilibrium tides that prescribes, for a fluid sphere, two high tides and two low tides everywhere on Earth. This obviously is far from what happens because of the irregular distribution of continents, the complicated ocean bathymetry, the declination of the Moon, among other factors. Through a mathematical derivation one can show that this tide-generating force between two celestial bodies is proportional to the product of their mass and inversely proportional to the cube, not the square, of their distance. Therefore, distance may matter much more than mass when it comes to producing tides. Considering that the Sun is 2.7×10^7 times more massive than the Moon but 390 times farther from Earth, the relative force from the Sun to produce tides on Earth is 0.46 that of the Moon. In other words, the Moon's tide-generating force on Earth is 2.17 times greater than the force from the Sun.

In addition to the astronomic tides described above, there are atmospheric tides, which are of thermodynamic origin. Atmospheric tides are manifested by oscillations in atmospheric pressure of 1–4 hPa at the Earth's surface (Figure 3.1). Heating of the atmosphere by the Sun causes barometric pressure variations at sea level with periods of 12 h and 24 h. These atmospheric tides are most relevant in

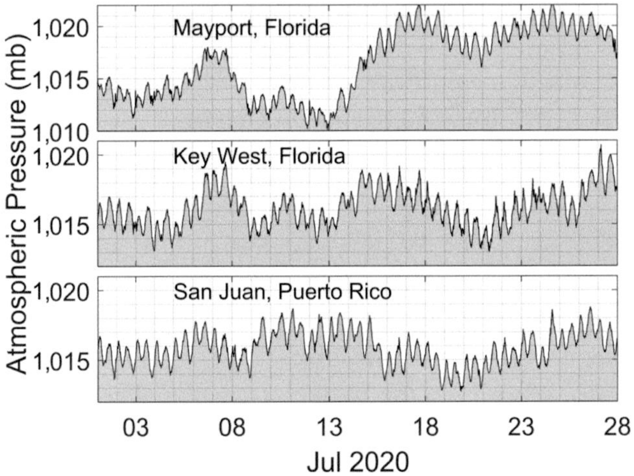

Figure 3.1 Atmospheric pressure (in millibars or hectoPascals) at three different sites. San Juan is in a tropical latitude (18° 27.6′ N) while the other two stations are in subtropical latitudes. Key West is actually close to the Tropic of Cancer at 24° 33′ N, and Mayport is at 30° 23.9′ N. All stations are influenced by semidiurnal S_2 atmospheric tides as the peaks and troughs occur at the same time each day. However, the influence is more notable at San Juan than at Mayport.

latitudes lower than 30°, where heating on Earth is highest. Although widely disregarded as irrelevant drivers in the coastal ocean, atmospheric tides should be scrutinized. They may cause sea-level variations at 12 and 24 h periodicities in tropical and subtropical coastal lagoons where the tide from the ocean (of astronomic origin) completely dissipates over shallow bathymetry or because of choking points. Such attenuation influence is described later in this chapter.

3.2 Main Harmonics, Constituents, or Harmonic Constituents

The dominant effects of the Moon and the Sun on Earth's tides may be represented as the superposition of oscillations with different periods or frequencies. Each oscillation with its own frequency may be referred to as a *tidal constituent*, *harmonic constituent*, or simply a *harmonic*. It follows that the tidal elevation η at any point on Earth and at any given time t can be represented as a sum of M harmonics (or sinusoid waves), as follows:

$$\begin{aligned}\eta = \sum_{k=1}^{M} A_k \sin(\omega_k t + \phi_k) = \\ A_{M2}\sin(\omega_{M2}t + \phi_{M2}) + A_{S2}\sin(\omega_{S2}t + \phi_{S2}) + A_{N2}\sin(\omega_{N2}t + \phi_{N2}) + \ldots\end{aligned} \tag{3.1}$$

where A is the amplitude of the sinusoid (or harmonic or tidal constituent) with angular frequency ω (= 2π/Period) and with a phase lag ϕ (in radians), relative to an arbitrary reference time, usually that of Greenwich. The tidal harmonics to which we must pay initial attention at a study site are listed in Table 3.1 with their respective period and angular frequency ω.

Table 3.1 includes three harmonics infrequently considered: the larger lunar elliptic semidiurnal (N_2), the solar semiannual (SSA), and the solar annual (SA). At least on the eastern coast of the United States, these three are quite influential in

Table 3.1. *Main harmonics associated with astronomic tides on Earth, showing their symbol, period in hours (h) or days (d), and angular frequency*

Tidal Constituent	Symbol	Period	Angular frequency
Principal lunar semidiurnal	M_2	12.4206 h	0.5059 rad/h
Principal solar semidiurnal	S_2	12.0000 h	0.5236 rad/h
Larger lunar elliptic semidiurnal	N_2	12.6583 h	0.4964 rad/h
Lunisolar diurnal	K_1	23.9345 h	0.2625 rad/h
Lunar diurnal	O_1	25.8193 h	0.2434 rad/h
Solar semiannual	SSA	182.628 d	0.0344 rad/d
Solar annual	SA	365.256 d	0.0172 rad/d

the amplitude of the tides and should be pondered in any analysis. The N_2 is responsible for consecutive inequalities in spring or neap tidal ranges within one month (Figure 3.2, Sewells Point). In areas where the amplitude of the $N_2 > S_2$, spring (or neap) tides will be different within practically any given month. This should be accounted for in sampling strategies. The semiannual and annual tides are responsible, for example, for the highest tides of the year. Coastal flooding caused exclusively by tides could be associated with the annual tidal harmonic, which would be responsible for what is sometimes referred to as "King Tides." Harmonics SSA and SA should also be considered in sampling schemes at any study site.

Even though equation 3.1 may look threatening, it only represents various waves interacting with each other. As an exercise to feel more comfortable with the expression, a reader should draw equation 3.1 ascribing arbitrary values to the amplitude and phase of different combinations of harmonics with angular frequencies consistent with those shown in Table 3.1.

Tides at a given place can be characterized on the basis of their "form factor," F, which is a ratio of tidal constituent amplitudes (A_k, equation 3.1). Specifically, F is the ratio between diurnal and semidiurnal harmonic amplitudes. Values of F have been traditionally calculated as

$$F = (A_{K1} + A_{O1})/(A_{M2} + A_{S2}) \tag{3.2}$$

and are the most basic means of characterizing tides at a place with known amplitudes of the corresponding tidal constituents. Following this calculation, we say that a tide is *semidiurnal* when $F < 0.25$, that is, when the sum of semidiurnal constituents M_2 and S_2 is at least four times greater than the sum of diurnal harmonics K_1 and O_1. This happens when consecutive high tides within the same day have nearly the same level (Figure 3.2). When the form factor falls in the range $0.25 < F < 1.25$, it is said that the tides are *mixed with semidiurnal dominance*, which occurs at sites where consecutive high tides within the same day have noticeably different levels (diurnal inequality, Figure 3.2). The factor falls in the interval $1.25 < F < 3$ where the tides are *mixed with diurnal dominance*, reflecting mostly one appreciable high tide per day. Semidiurnal tides may appear during periods of equatorial tides (Section 3.3, Figure 3.2). Finally, sites with $F > 3$ indicate *diurnal* tides (Figure 3.2).

This approach works only for the daily character of the tide and does not consider the potential influence of N_2, which could be larger than S_2. In some parts of the world where $N_2 > S_2$, such as the Atlantic coast of the USA, values of F should be reported to include the amplitude of N_2, instead of S_2, while specifying the nuance of the semidiurnal harmonics. In some instances, this distinction will make a negligible difference in F but not in others.

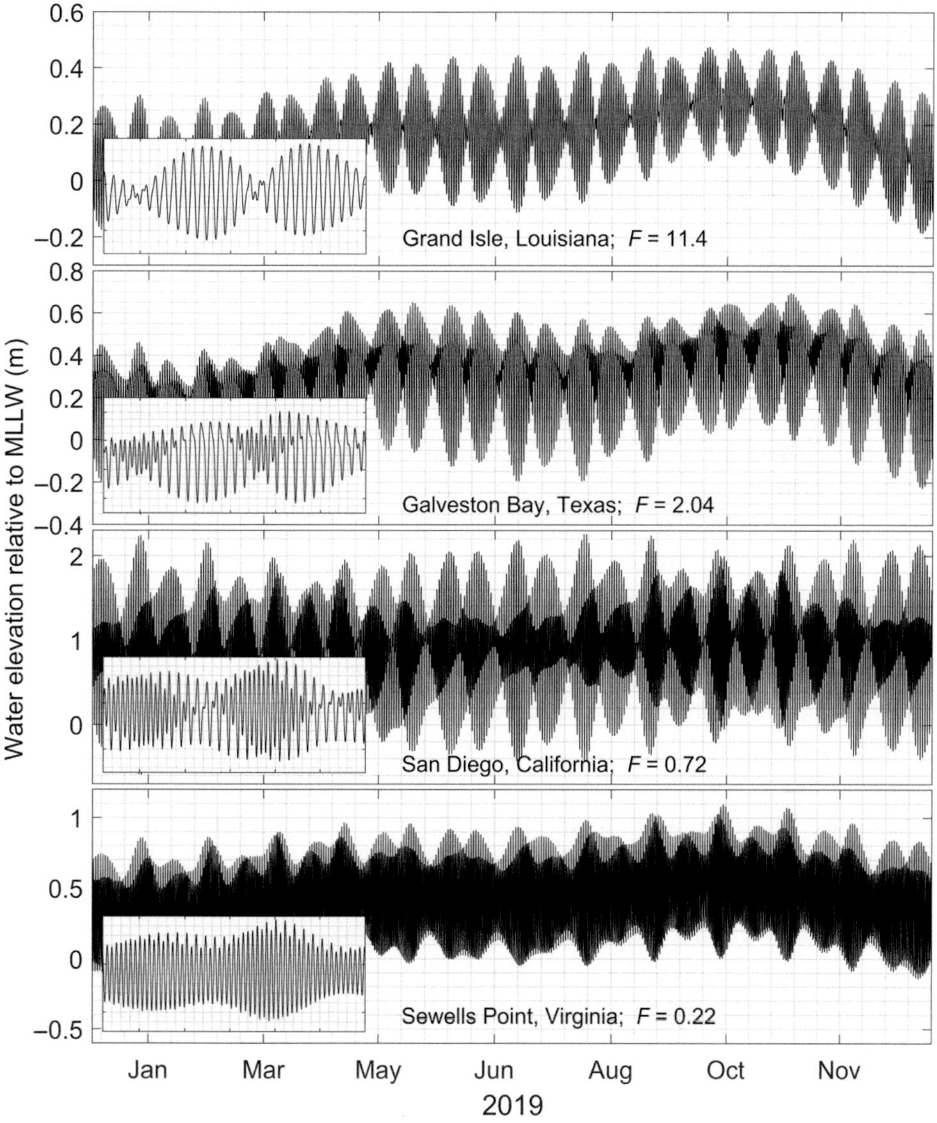

Figure 3.2 One year of predicted tidal records for each of the four types of tides, according to their form factor F. Each record also shows a close-up for April. In the close-ups, Sewells Point exhibits semidiurnal tides with relatively small diurnal inequality (difference between consecutive low tides) and synodic fortnightly modulation. The form factor at this location is 0.21 if the N_2 is considered instead of the S_2. San Diego displays two highs and two lows per day with marked diurnal inequality. Galveston Bay illustrates mostly one tide per day, with some asymmetry between high and low tides because of the semidiurnal influence. Also, the semidiurnal influence in Galveston becomes most evident at equatorial tides (periods of smallest range). Grand Isle describes one high and one low per day with a well-defined fortnightly (declinational) modulation.

In addition to the form factor *F*, tides can be distinguished according to their range, which is the difference in meters between high tide and low tide. The range is taken for the total tide, that is, the sum of all harmonics. Thus, *microtidal* regimes feature tidal ranges <2 m, as in all of the Gulf of Mexico and most southeastern coast of the USA. *Mesotidal* coastal regions display ranges between 2 and 4 m, as in most of the Gulf of Alaska. Finally, *macrotidal* coasts are influenced by tidal ranges >4 m, as in the Bay of Fundy or Hudson Bay in Canada.

3.3 Spring and Neap Tides: Tropic and Equatorial Tides

It is widely acknowledged that the tides with the largest tidal range at a site result from the overlap of the gravitational attraction from the Sun and the Moon. These periods of highest tidal range in one month are known as *spring tides*. In contrast, periods when the gravitational attraction from the Moon is orthogonal, or perpendicular, to that of the Sun result in the smallest ranges of a month and are called *neap tides*. The period between consecutive spring or neap tides is known as a *synodic fortnight* and equals 14.77 d. This period represents, for example, the time from New Moon to Full Moon and equals the modulation, or interference or difference in frequency, between the M_2 and S_2 harmonics, that is, 1/(1/12 h – 1/12.4206 h). The modulation period of 14.765 d (see Animation 3.1) is usually referred to simply as *fortnight*. However, it should be referred to as the *synodic fortnight*. Twice the period of the synodic fortnight, that is, 29.53 d, represents the time between consecutive Full (or New) Moons and is the *synodic month*. Thus, we must make the important distinction between the *synodic fortnight* and the actual *fortnight* or *declinational fortnight*. The latter has to do with the position of the Moon relative to Earth's Equator or Tropics, that is, the fortnight has to do with the Moon's declination.

At maximum declination, the Moon is over the Tropics (*Tropic* tides – analogous to spring tides). At minimum declination, the Moon is over the Equator (*equatorial* tides – analogous to neap tides). The excursion of the Moon from its declination over the Tropic of Capricorn to the Tropic of Cancer takes 13.66 d, one fortnight. Such period equals the modulation or interference between the diurnal harmonics K_1 and O1, that is, 1/(1/23.93447 h – 1/.25.81934 h). Twice that period is the period between consecutive declinational maxima over the same tropic, or 27.32 d. This is the *sidereal month*. Therefore, the type of tidal modulations over one month (or half of one month) will be determined by the dominance of semidiurnal or diurnal tides.

Coastal areas dominated by semidiurnal tides (M_2 & S_2, or $F < 0.25$) will feature *synodic fortnightly* variability ($\approx$14.77 d) between consecutive spring or neap tides (Figure 3.3). One should be aware, however, of the possible influence of the N_2 harmonic, which has to do with the ellipticity of the Moon's orbit. In such

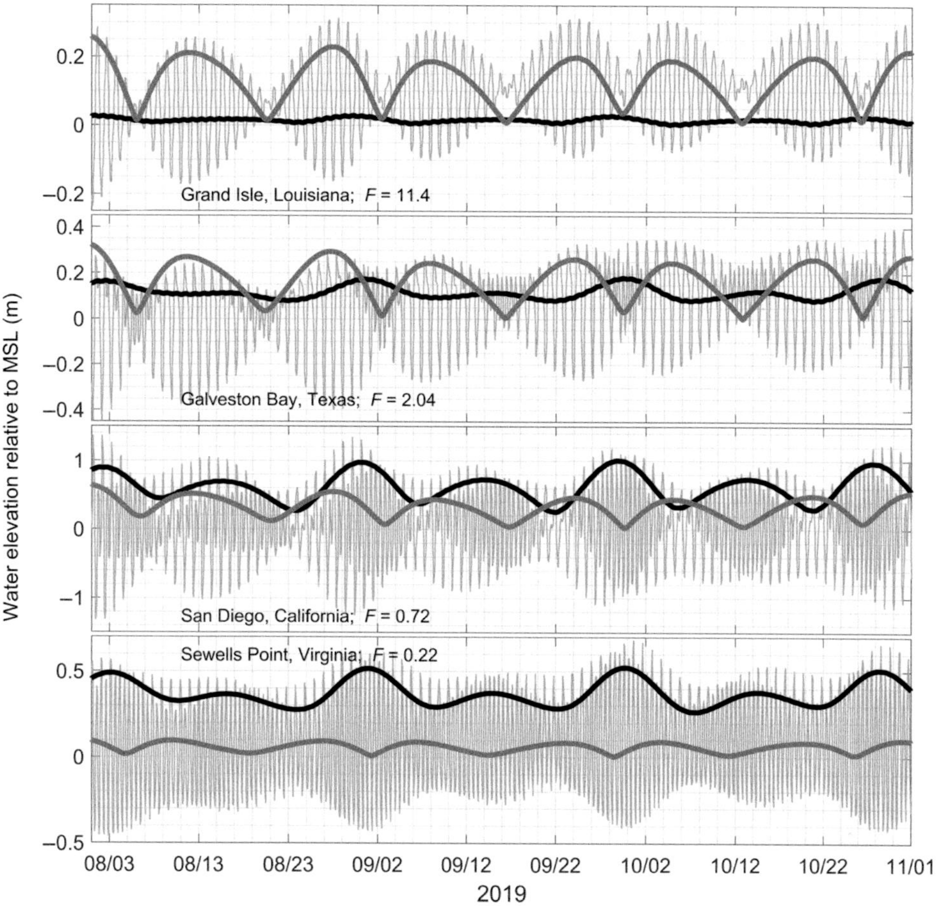

Figure 3.3 Fortnightly modulations for different types of tides. Semidiurnal tides (Sewells Point) show dominance of semidiurnal amplitude modulation (thick black line) compared to diurnal amplitude modulations (thick gray line). Thin gray lines display three months of records at the same sites of Figure 3.2. Diurnal amplitude importance (thick gray line) increases from the bottom to the top, showing coincidence of maximum diurnal tidal ranges with maximum modulation of this harmonic. The same coincidence occurs for the semidiurnal ranges.

elliptical orbit, the Moon's position closest to Earth is the *perigee* and its farthest distance is the *apogee*. The time from perigee to perigee, or apogee to apogee, is the *anomalistic month*, which equals 27.55 d. In coastal areas where $N_2 > S_2$, there will be this 27.55 d modulation, which is also the difference in frequencies between N_2 and M_2 (see Sewells Point, Figure 3.2). As mentioned before, those areas will feature consecutive spring tides, or neap tides, that are asymmetric (see thick black line for Sewells Point in Figure 3.3).

In addition, coastal areas dominated by diurnal harmonic constituents ($F > 3$) will display fortnightly variability (13.66 d) between consecutive maxima or minima in tidal range (see thick gray line for Grand Isle in Figure 3.3). Regions with mixed tides and semidiurnal dominance will exhibit tidal range modulations influenced by synodic fortnights and declinational fortnights. Extreme diurnal inequalities will appear at times of greatest diurnal amplitude modulation (see San Diego in Figure 3.3). Accordingly, mixed tides with diurnal dominance regimes will feature increased influence of semidiurnal tides at the peak of the synodic fortnight and enhanced diurnal tides at the height of the declinational fortnight (see Galveston Bay, Figure 3.3). Note in Figure 3.3 that semidiurnal amplitude modulations (thick black line) are in phase at the four sites shown as they are linked to the phases of the Moon. Similarly, the diurnal amplitude modulations (thick gray line) also occur at the same time as they are related to the Moon's declination relative to Earth's Equator.

3.4 Tidal Waves in Semienclosed Basins

This section deals with basic quantitative approaches to study tidal propagation in semienclosed basins. (Refer to Chapter 2, equation 2.40, to remember the origin of the quantitative expressions of this section.) A brief reminder is that the relationships presented here are anchored in Newton's Second Law. The physics describing the phenomena of this section are relatively straightforward: you have a driving force per unit mass that causes accelerations, typically a water-level slope; occasionally you will have friction affecting the accelerations or Earth's rotation (*Coriolis* effects).

The first type of generic wave described is a frictionless, or inviscid, wave. This type of wave, a *plane wave*, is treated afterward to include friction effects and the influence of coastline convergence. The frictionless wave is then treated under the effects of Earth's rotation. The section closes with a treatment of tidal current structure under the influence of laterally varying bathymetry.

3.4.1 Frictionless Tidal Wave

In this type of wave, fluid accelerations ($\partial U/\partial t$) along a semienclosed basin are produced by a pressure gradient force (per unit mass) represented by water-level slopes ($\partial \eta/\partial x$). In symbolic representation, such expression may be written as:

$$\underbrace{\frac{\partial U}{\partial t}}_{\textit{fluid accelerations}} = \underbrace{-g\frac{\partial \eta}{\partial x}}_{\textit{pressure gradient}} \tag{3.3}$$

where x is the along-basin direction, t is time, η is surface elevation, U is the along-basin tidal (or in general, wave) current averaged over the mean water column depth H, and g is the gravitational field force per unit mass, also known as acceleration caused by gravity. Equation 3.3 is also a simplification of the general form of the *along-basin momentum balance* assuming linear motion, without rotation effects or friction, in a homogeneous fluid.

Equation 3.3 is beautifully simple in the sense that it is analogous to many oscillatory motions, such as that of a frictionless pendulum or a frictionless spring or an infinitely vibrating string. All these motions are perturbed initially and oscillate without stopping. Equation 3.3 has a solution for η or U, but first we need to establish a link between these two variables. In most problems of fluid mechanics, we need to complement the conservation of momentum with an expression for conservation of mass. In this case, conservation of mass, also known as *continuity*, provides the linkage needed:

$$\frac{\partial U}{\partial x} = -\frac{1}{H}\frac{\partial \eta}{\partial t}. \tag{3.4}$$

This is equivalent to equation 2.11, but integrated vertically and neglecting variations in the across-basin direction y. We can combine conservation of momentum and conservation of mass (equations 3.3 and 3.4) to obtain

$$\frac{\partial^2 \eta}{\partial t^2} = gH\frac{\partial^2 \eta}{\partial x^2} = C^2\frac{\partial^2 \eta}{\partial x^2}, \tag{3.5}$$

where C equals $\sqrt{gH}$ and indicates the speed at which a tidal wave form moves. For those interested, equation 3.5 is a linear, hyperbolic partial differential equation obtained in one of two ways: (i) by differentiating equation 3.4 with respect to time, which provides an expression for $\partial U/\partial t$ that in turn is substituted in equation 3.3 or (ii) differentiating equation 3.3 with respect to x, which furnishes a representation for $\partial U/\partial x$ that is replaced in equation 3.4. Most relevant, regardless of its derivation, is that equation 3.5 represents the *wave equation* in one dimension. It provides the behavior of the water level η (the dependent variable) as a function of time t and along-wave or along-basin direction x. Remember that this wave equation arose from basic physics of an accelerated motion driven by an initial perturbation; in this case a water-level slope $\partial\eta/\partial x$.

The wave equation (3.5) has several possible solutions. One of them, which is adopted here, is called the d'Alembert's solution represented by a sinusoidal wave form for η. We call this a plane wave:

$$\eta = A\cos\left(\frac{2\pi}{\lambda}x - \frac{2\pi}{T}t\right) = A\cos(\kappa x - \omega t) \tag{3.6}$$

where A is the wave amplitude, λ is its wavelength or distance between consecutive crests, T is its period or time between successive crests (or any reference point on the wave). The angular frequency ω is inversely related to T by $\omega = 2\pi/T$. The wave number κ is analogous to the wave angular frequency, but in space, and relates to λ by $\kappa = 2\pi/\lambda$. We can think of κ as the extent of a wave, that is, a relatively large κ value signifies a relatively short wave, and vice versa. Equation 3.6 ensures non-zero amplitude at the entrance to a basin, that is, at $x = 0$.

Visualize the animation of a wave changing in space and time and run the program (Animation 3.2).

An expression for the depth-averaged wave orbital velocity U, that is, the tidal velocity in this case, can be obtained by taking the expression for the plane wave η (equation 3.6), differentiating it with respect to time t and inserting the value of $\partial\eta/\partial t$ into the mass balance equation (equation 3.4), to integrate then with respect to x, to find

$$U = \frac{1}{H}\frac{\omega}{\kappa}A\cos(\kappa x - \omega t) = \frac{C}{H}A\cos(\kappa x - \omega t). \tag{3.7}$$

In this plane wave, the amplitude of the oscillatory velocity and the amplitude A of the water level fluctuation η are related by the factor C/H. Knowing the tidal amplitude and the depth of the water column we can guess the amplitude or strength of the tidal current for a frictionless tide. This can sometimes work in a real situation, but in coastal waters tidal oscillations are typically affected by frictional effects, as described in Section 3.4.2. Furthermore, equations 3.6 and 3.7 indicate that η and U are in phase because they both depend on $\cos(\kappa x - \omega t)$ as is the case in any *progressive* or *travelling wave* (see Figure 3.4).

Another possible solution to the wave equation (3.5) is the combination of waves traveling in opposite directions, to the right $A\cos(\kappa x - \omega t)$ and to the left $A\cos(\kappa x + \omega t)$:

$$\eta_s = A\cos(\kappa x - \omega t) + A\cos(\kappa x + \omega t) = 2A\cos(\kappa x)\cos(\omega t). \tag{3.8}$$

The expression on the rightmost portion, after the second equal sign, arises from a trigonometric identity (sum to product) involving a sum of cosines. This is a good opportunity to visualize the animation provided and apply the Matlab© code (Animation 3.3) to represent equation 3.8 and its *standing wave* behavior. Examination of equation 3.8 reveals that at any time t, the condition for η_s to be zero is when $\cos(\kappa x)$ equals 0, that is, where κx is $\pi/2$, $3\pi/2$, ... or where x equals $\lambda/4$, $3\lambda/4$, ... These positions are the *nodes* of the standing wave. On the other hand, the condition for the absolute value of η_s to be $2A$ is where $\cos(\kappa x)$ equals 1, that is, where κx is 0, π, 2π ... or where x equals 0, $\lambda/2$, λ, ... These are the *antinodes*.

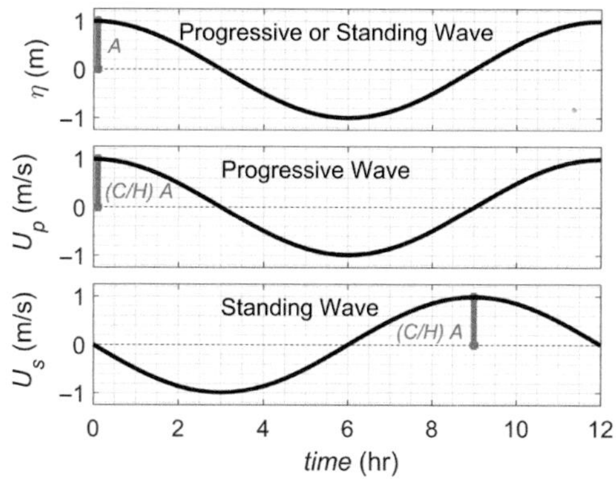

Figure 3.4 Relationship between η and U for progressive (in phase) and standing (in quadrature) waves.

As we did for the progressive wave, we can derive the oscillatory velocities U_s for a standing wave by differentiating equation 3.8 in time and then using continuity (equation 3.4) and integrating in x. Alternatively, we can differentiate 3.8 with respect to x and use the momentum balance (equation 3.7) to integrate in t. Either way, the result is:

$$U_s = \frac{2}{H}\frac{\omega}{\kappa}A\sin(\kappa x)\sin(\omega t) = \frac{2C}{H}A\sin(\kappa x)\sin(\omega t). \tag{3.9}$$

It is evident that elevations η_s and velocities U_s are in quadrature, that is, are 90° out of phase (see Figure 3.4) because the trigonometric functions in x and t are switched (compare equations 3.8 and 3.9). Maximum oscillatory velocities in a standing wave then occur at $\sin(\kappa x)$ equals $\pi/2$, $3\pi/2$, . . ., that is, where x is $\lambda/4$, $3\lambda/4$, . . ., that is, at the region (or time) with maximum change in η_s, that is, at the nodes. Oscillatory velocities in a standing wave are zero at the antinodes.

In coastal water bodies where the tide behaves like or close to a standing wave, maximum tidal currents occur between high tide and low tide, or between low and high tide. At these sites, slack periods (near-zero currents) occur at high or low tide. Keep in mind that this description applies to frictionless waves but tidal waves in coastal basins tend to be affected by friction. This will be explored further in Sections 3.4.2 and 3.4.3.

A standing wave may amplify in semienclosed basins that resonate with the wave (Figure 3.5). This resonance phenomenon depends on the basin's dimensions. We want to investigate the conditions that cause amplification of a wave in a semienclosed basin, after it enters at $x = L$.

If the wave oscillating at the entrance to the basin is written as

$$\eta = A_0\cos(\omega t) \tag{3.10}$$

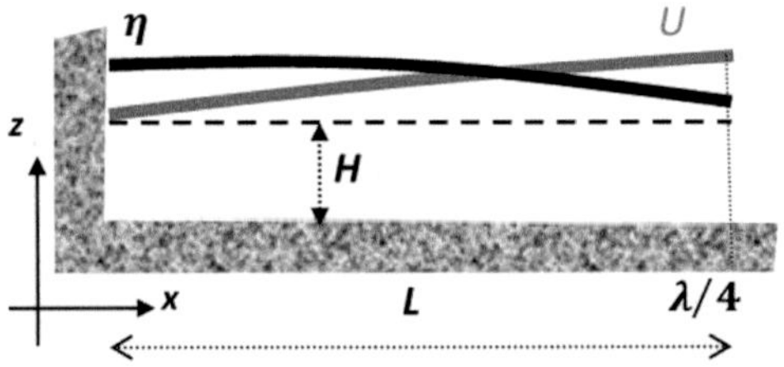

Figure 3.5 Longitudinal section in a semienclosed basin showing conditions for resonance in a frictionless wave. This is known as a quarter-wave resonator, where η is maximum at the head of the basin and U is greatest at its entrance.

and we substitute equation 3.10 into equation 3.8 at $x = L$, we get

$$2A\cos(\kappa L)\cos(\omega t) = A_0\cos(\omega t). \tag{3.11}$$

From expression 3.11 we can derive the amplitude increase in the basin, relative to the amplitude at the entrance:

$$A = \frac{A_0}{2\cos(\kappa L)}. \tag{3.12}$$

Values of A in equation 3.12 could exhibit unconstrained amplification inside the basin when the denominator tends to zero. This would be analogous to a sound wave resonating (or echoing) in a chamber, or the vibration of a guitar string resonating in a box (the guitar body), or a tuning fork on a resonance box. The condition for this amplification, or resonance, to happen is when κL tends to $\pi/2$, $3\pi/2$, $5\pi/2$, ..., that is, when $\cos\kappa L$ tends to zero, or $\kappa L = \underbrace{\frac{2\pi}{\lambda}}_{\kappa} L = -\frac{\pi}{2} + 2n\frac{\pi}{2}$,

where $n = 1, 2, 3, \ldots$ represents the "nth mode" of resonance, and n also indicates the number of nodes found in the basin of length L. This expression can be simplified to $L = \frac{\lambda}{4}[2n - 1]$, for $n = 1, 2, 3, \ldots$, which describes the length of the basin required for resonance. In other words, this expression indicates that a semienclosed basin should at least fit one-quarter of a standing wave to resonate. We call this a quarter-wave resonator. But for a wave like a tidal wave, its wavelength equals its celerity $C(=\sqrt{gH})$ times its period T, that is, $\lambda = CT$ (distance times time), so that, $L = \frac{CT}{4}[2n - 1]$. Finally, the period T needed for resonance to ensue, which we will denote as T_N is

$$T_N = \frac{4L}{C}[2n - 1]. \tag{3.13}$$

This is the version of *Merian's formula* for semienclosed basins and represents *the natural period of oscillation for resonance*, that is, the period at which we must

Table 3.2. *Mean properties used in Merian's formula to calculate T_N (equation 3.13) for $n = 1$*

Estuary	H (m)	L (km)	C (m/s)	T_N (h)
Long Island Sound	20	180	14	14
Chesapeake Bay	10	250	10	28
Bay of Fundy	70	250	26	10.7

force the basin to display resonance. Another version of Merian's formula, the original, applies to closed basins such as lakes, where it was first proposed. That version has a 2 instead of a 4 in the numerator of equation 3.13, representing one half wavelength of a resonating standing wave, also known as *seiche*. Most frequently, in semienclosed basins we use the ratio $4L/C$ (mode 1, $n = 1$) to determine whether an estuary can resonate to the tide. There are many examples of the use of Merian's formula in estuaries. Three of them are illustrated in Table 3.2.

It is evident that two out of the three basins in Table 3.2, Long Island Sound and the Bay of Fundy, could possibly resonate to semidiurnal tidal forcing as their natural period is close to 12 hours. Indeed, these two systems exhibit amplification of the tidal wave inside the basin. The Bay of Fundy features the largest tidal range on Earth, near 12 m (close to a four-story building!) at Minas Basin. Resonance in these two basins is not only caused by T_N but also by their convergent coastlines that concentrate the wave energy and make it grow. Coastline convergence effects are described in Section 3.4.3. Chesapeake Bay has a period of oscillation far removed from its dominant tidal period of ~12 h and therefore is non-resonant to tidal forcing.

3.4.2 *Effects of Friction on a Tidal Wave*

In shallow semienclosed basins, it is most likely that tidal waves will be affected by friction. The momentum or dynamic balance that describes this situation is analogous to that of any damped oscillation or damped harmonic oscillator such as a spring or a pendulum oscillation. For a progressive tidal wave, we can write such momentum balance for a depth-averaged flow U as

$$\underbrace{\frac{\partial U}{\partial t}}_{\textit{acceleration}} = \underbrace{-g\frac{\partial \eta}{\partial x}}_{\textit{pressure gradient}} \underbrace{-\frac{C_b u_b |u_b|}{H}}_{\textit{bottom friction}} \tag{3.14}$$

where C_b is an adimensional bottom drag coefficient, with canonical value of 0.0025 (although it may vary), and u_b is the near-bottom velocity (see also equation 2.42).

Equation 3.14 has the same elements as the dynamical representation for a plain wave (equation 3.3) plus bottom friction. Thus, equation 3.14 symbolizes an attenuated plane wave, or in this case, an attenuated tidal wave. If we represent the near-bottom velocity as an oscillatory motion with amplitude U_0, that is, $u_b = U_0 \cos(\omega t)$, and expand the bottom friction term in a Fourier series, we obtain, to the lowest order

$$\frac{\partial U}{\partial t} = -g\frac{\partial \eta}{\partial x} - \underbrace{\frac{8}{3\pi}\frac{C_b U_0}{H}}_{r} U = -g\frac{\partial \eta}{\partial x} - rU. \tag{3.15}$$

The coefficient r is a linearized bottom friction coefficient (units of s^{-1}) that converts the nonlinear bottom friction term (proportional to u_b^2, equation 3.14) to a linear term, as seen after the second equal sign in equation 3.15. This equation has the general form of the momentum balance for any damped harmonic oscillator. It is a homogeneous differential equation with a solution that depicts an initial perturbation decaying exponentially with time. The mass balance, or continuity equation, that accompanies the momentum balance (3.15) is the same as for the plane wave (equation 3.4).

A solution that satisfies 3.15 has a relatively simple form that still depends on x and t:

$$\eta_a = Ae^{-\mu x}\cos(\omega t - \kappa x) \tag{3.16}$$

where μ is an attenuation coefficient that equals $r/[2\sqrt{gH}]$. Equation 3.16 describes a wave of amplitude A at the entrance to a basin ($x = 0$) decaying exponentially along the distance x in the basin. The solution is consistent with that of the plane wave, but with exponential decay along its path.

Applying the continuity equation, we obtain an expression for the depth-averaged tidal current attenuated in x:

$$U_a = \frac{A\omega}{H\sqrt{\mu^2 + \kappa^2}}e^{-\mu x}\cos(\omega t - \kappa x + \alpha,) \tag{3.17}$$

where $\alpha = \text{atan}[r/2\omega]$ is a phase lag between η_a and U_a caused by friction. Extremes in U_a precede those of η_a, which would mean that maximum flood occurs some time $[\alpha \times 12/2\pi \text{ hours}]$ before high tide. At this point, it would be relevant to visualize an animation of equations 3.16 and 3.17 or run the Matlab© code (Animation 3.4) to understand the behavior of η_a and U_a better.

Equations 3.16 and 3.17 portray an attenuated *progressive wave*. Frictional effects can also attenuate a *standing wave*. In that case, the wave behavior would be illustrated by:

$$\eta_a = Ae^{-\mu x}[\cos(\omega t - \kappa x) + \cos(\omega t + \kappa x)] \tag{3.18}$$

and

$$U_a = \frac{A\omega}{H\sqrt{\mu^2 + \kappa^2}} e^{-\mu x}\big[\cos(\omega t + \kappa x + \alpha) \\ - \cos(\omega t - \kappa x - \alpha)\big].$$

As with the frictionless standing wave, the attenuated standing wave shows near quadrature between water level and flow. As an exercise, the reader could try to visualize the behavior of equations 3.18 by modifying the code provided for an attenuated progressive wave, which draws the animation of equations 3.16 and 3.17. An example is provided without the code.

3.4.3 Effects of Coastline Convergence

In Section 3.4.1, we explored a frictionless tidal wave moving along a channel with straight and parallel coastlines. In Section 3.4.2 we presented the motions of an attenuated tidal wave under the same geometry. In this section, we alter the geometry slightly by including coastlines that converge toward the head of the basin. We restrict our presentation to a conceptual assessment of the effects of coastal convergence. A rigorous treatment requires inclusion of nonlinear distortions to the tide, which will be described in Chapter 4.

Coastline convergence effects can be quantified in the conservation of mass, or continuity, expression (not shown explicitly here) as the ratio between local half width $B(x)$ versus a *convergence length scale* (L_c), in such a way that the two symmetric coastlines are distributed along the basin as

$$B(x) = B_e e^{-x/L_c}, \tag{3.19}$$

where B_e is the half width of the basin at its entrance (Figure 3.6). The convergence ratio ($\Gamma = B(x)/L_c$) in most systems decreases from the entrance to the head of the basin. Comparing different basins, the greater the ratio, the more convergent their coastlines. Thus, the tide will be attenuated by friction or will be amplified, under frictionless or nearly frictionless conditions, by coastline convergence or by resonance.

Ultimately, the competition between friction and coastline convergence will determine one of three outcomes of the tidal amplitude relative to that at the basin's entrance. The first outcome is that the tidal amplitude remains nearly constant along the basin because of frictional effects being nearly equal to coastline convergence influence. In this case, the tidal wave is said to be *synchronic*. The second outcome is that the tidal wave amplifies because coastline convergence is greater than frictional effects, representing a *hypersynchronic*

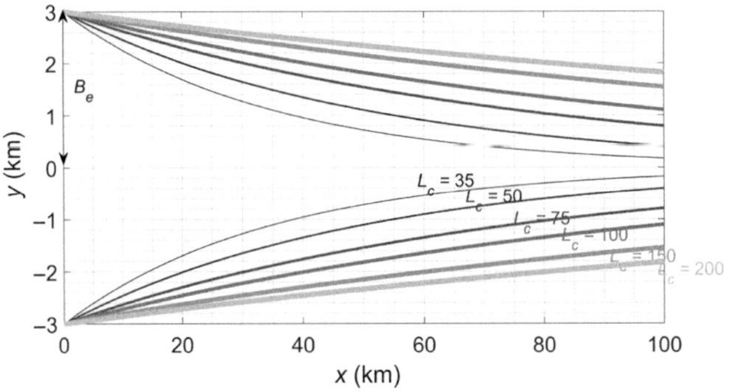

Figure 3.6 Illustration of equation 3.19 displaying convergent channels with different convergence length scales L_c.

basin. The third outcome is tidal wave attenuation because of frictional effects being greater than coastline convergence, resulting in a *hyposynchronic* basin.

Frictional effects on tidal waves can be determined by scaling the ratio between friction (or stress divergence in the momentum balance) $(A_z u/H^2)$ and local accelerations $(u\omega)$, which yields a nondimensional number that could be referred to as Stokes number. This ratio is $\delta = A_z/\omega H^2$ and represents the proportional depth of frictional influence, analogous to the actual frictional depth obtained in Stokes' second problem in fluid mechanics. From equation 3.15, δ can also be expressed in terms of the linear friction coefficient as r/ω, with the same physical interpretation. Thus, when $\delta < 0.1$ the depth of frictional influence extends to $<10\%$ of the water column, that is, nearly frictionless conditions.

With δ and Γ (the convergence ratio = $B(x)/L_c$), we can find four extreme combinations portraying distinct scenarios. The first combination involves low δ and low Γ, representing nearly frictionless basins with weak-to-no coastline convergence. That case is like the one considered in Section 3.4.1 in which the tidal wave will be synchronous under no resonance or could also be hypersynchronous if there is resonance. Examples of this combination are found in some fjords. The second combination relates to low δ but high Γ. Under these circumstances, the mass balance described in Section 3.4.1 would have to be modified to accommodate convergence in basins' width, which will likely result in hypersynchronous tides. Examples of this combination are found in deep, convergent basins like the Bay of Fundy and some fjords. The third and fourth combinations feature δ of order 1, indicating dynamical effects of friction. Under low Γ, the tidal wave attenuates and displays hyposynchronic conditions. Most shallow estuaries fall in this category. The situation of high δ and Γ will produce a scenario that may result in any of the three synchronic conditions depending on whether friction or coastline convergence prevails in modifying the tide. Several estuaries with convergent coastlines in the UK fall in this combination of non-negligible Γ relative to δ.

3.4.4 Effects of Earth's Rotation on a Frictionless Tidal Wave

We return to the plane wave (Section 3.4.1) and explore the effects of *Coriolis* acceleration on the behavior of η and U. Assuming that the wave is propagating along a coast in the x direction and that the orbital velocities (or tidal currents) perpendicular to this boundary vanish, that is, $V = 0$, the momentum balance in x is the same as for a plane wave. In the lateral direction y, the momentum balance is between a lateral pressure gradient (lateral surface slopes) and *Coriolis* acceleration. In other words, the y momentum balance is *geostrophic*:

$$\begin{aligned}
\underbrace{\frac{\partial U}{\partial t}}_{\textit{fluid accelerations}} &= \underbrace{-g\frac{\partial \eta}{\partial x}}_{\textit{pressure gradient}} \\
\underbrace{fU}_{\textit{Coriolis accelerations}} &= \underbrace{-g\frac{\partial \eta}{\partial y}}_{\textit{pressure gradient}}
\end{aligned} \tag{3.20}$$

where f is the Coriolis parameter (equations 2.33 and 2.34). Conservation of mass is the same as equation 3.4 because there is no normal flow at the boundary ($V = 0$). The solution of equations 3.20 and 3.4 is

$$\eta = Ae^{-\frac{y}{R}} \cos(\kappa x - \omega t), \tag{3.21}$$

where R equals $\sqrt{gH}/f$ $(= C/f)$, which has units of length. The scale R denotes the Rossby radius of deformation, which represents a length scale over which Earth's rotation effects are influential in the transverse structure of the wave. Combining equation 3.21 with 3.4 yields

$$U = A\frac{C}{H}e^{-\frac{y}{R}} \cos(\kappa x - \omega t). \tag{3.22}$$

Both η and U have the same shape as a plane wave when disregarding the exponential term ($e^{-\frac{y}{R}}$). This term provides an exponential decay in both variables in the y direction, that is, as we move away from the boundary along x. In other words, the water elevation and flow in this wave decrease exponentially, in the transverse direction, over a scale R. Values of R are typically between order 10 and 100 km. The wave described by equations 3.21 and 3.22 is known as a coastal Kelvin wave and can be visualized better in the animation provided and by running the Matlab© code furnished (Animation 3.5). This wave also has the form of the frictional wave, or attenuated wave (equations 3.16 and 3.17). However, the Kelvin wave decays in the direction that is perpendicular to the wave propagation, instead of along the same direction, as in the attenuated wave.

As described in equations 3.8 and 3.9 for plane standing waves, the points of destructive interference for two plane waves traveling in opposite directions are the *nodes*. Within a semienclosed basin that is wide enough to allow ingoing and outgoing Kelvin-type waves, or plane waves influenced by Earth's rotation, there will be areas of destructive interference. Instead of nodes, there will be regions where *tidal amplitude is minimum (or zero)* and where *phase lines converge*. These regions are called *amphidromic points*, which can be visualized in Animation 3.6 and by running the code provided. In a semienclosed basin, amphidromic points exhibit the lowest amplitude for semidiurnal or diurnal harmonics at a region denoted by a circular contour (Figure 3.7). The convergence of phase lines toward the center of the amphidromic region, that is, toward the amphidromic point, indicates that the specific harmonic wave rotates around this point. The sense of rotation is counterclockwise in the northern hemisphere and clockwise in the southern hemisphere.

Amphidromic points are observed for semidiurnal tides in several gulfs like the Red Sea, the Persian Gulf, the Gulf of Saint Lawrence, and the Gulf of California.

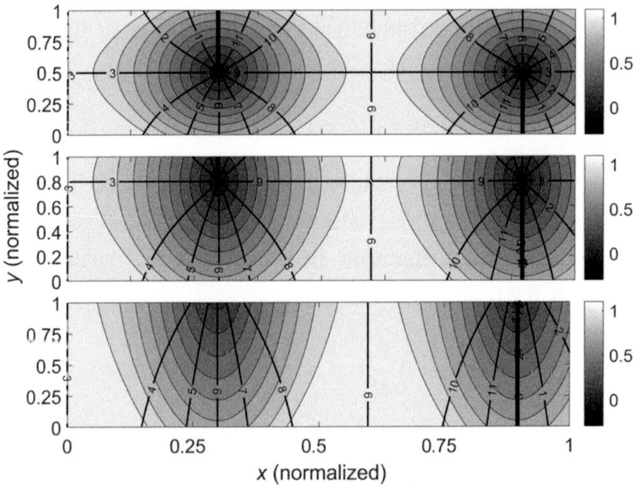

Figure 3.7 Idealized and conceptual co-tidal charts for a conjunction of two opposing Kelvin waves under different friction conditions in the northern hemisphere. In all charts, shaded contours indicate co-amplitude lines associated with the color bars and the labeled contours represent co-phase lines in hours. All axes have been normalized. The upper panel illustrates frictionless conditions, the middle panel represents intermediate friction, and the lower panel shows highly frictional conditions.

These are semienclosed basins that are wide enough (~200 km, actually two Rossby radii wide) to nearly fit Kelvin waves along the two boundaries that have roughly the same orientation. When a basin is narrower than two Rossby radii, amphidromic points of semidiurnal or diurnal tidal harmonics may appear closer to one of the along-basin boundaries than the other (Figure 3.7). In this case, the minimum in harmonic amplitude is still identifiable as a nearly circular contour. Also distinguishable is the convergence of phase lines toward the center of that contour region. In narrower basins, still tens of km wide, the semienclosed basin will exhibit an amplitude minimum and a convergence of phase lines toward one of the coasts. In this case, there will be no closed contour of minimum harmonic amplitude. This is a *virtual amphidromic point* (Figure 3.7). Virtual amphidromic points can be explained dynamically from the inclusion of frictional effects in the basin, in such a way that the outgoing Kelvin wave is attenuated relative to the incoming wave (Figure 3.7). Even though these basins are non-resonant to tidal forcing, they display Earth's rotation effects through these virtual amphidromic points. Examples of these systems with virtual amphidromes are found in Chesapeake Bay and the Bay of Fundy.

3.4.5 Effects of Lateral Bathymetry on Tidal Flows

This section has a little more advanced quantitative concepts, relative to previous sections in this chapter. However, it describes fundamental behaviors in tidal currents, which are greatly affected by bathymetric variations across and along a basin. When the dynamics of a semienclosed basin are dominated by friction, the maximum currents will appear over the deepest part of the cross-section. One can compare measurements of tidal flow amplitudes across a channel to those expected from simplified dynamics. By scaling the dominant forces related to tidal currents, and *assuming along-basin uniformity*, the momentum and mass balances may be written as

$$\begin{aligned} \frac{\partial u_0}{\partial t} &= -g\frac{\partial \eta_0}{\partial x} + A_z\frac{\partial^2 u_0}{\partial z^2} \\ f u_0 &= -g\frac{\partial \eta_0}{\partial y} + A_z\frac{\partial^2 v_0}{\partial z^2} \\ \frac{\partial v_0}{\partial y} &+ \frac{\partial w_0}{\partial z} = 0 \end{aligned} \tag{3.23}$$

In equations 3.23, the subindices 0 indicate tidal properties. These balances provide tidal flow fields (u_0, v_0, w_0) as a function of time *t*, cross-channel direction *y*, and depth *z*. Equations 3.23 form a system of partial differential equations that has an exact solution. Briefly, the system can be solved prescribing oscillatory behavior for $(u_0, v_0, w_0, \partial\eta_0/\partial x, \& \ \partial\eta_0/\partial y)$. The solution for u_0 is (where the hat denotes a complex number variable)

$$\hat{u}_0 = \frac{ig}{\omega}\frac{\partial\hat{\eta}_0}{\partial x}\left[1 - \frac{\cos h(\alpha_z z)}{\cos h(\alpha_z H)}\right]. \tag{3.24}$$

In equation 3.24, the imaginary number $i = \sqrt{-1}$ indicates a complex variable that provides amplitude and phase to the along-basin tidal current and along-basin water level slope $\partial\hat{\eta}_0/\partial x$, g is gravity, ω is the tidal frequency, α_z is a frictional length scale that equals $\sqrt{i\omega/A_z}$ analogous to the scale derived from Stokes' second problem in fluid mechanics. Moreover, $H(y)$ describes the cross-section depth profile, which can be arbitrary. In 3.24, the along-basin water-level slope $\partial\hat{\eta}_0/\partial x$ is estimated with the cross-sectional average of the tidal current amplitude $\overline{U_0}$ and the cross-sectional area A_s, in addition to the other variables appearing in 3.24:

$$\frac{\partial\hat{\eta}_0}{\partial x} = \frac{\overline{U_0}\omega A_s}{ig}\left[\int_0^{2B}\frac{1}{\alpha}\{\alpha_z H - \tan h(\alpha_z H)\}dy\right]^{-1}. \tag{3.25}$$

With equations 3.24 and 3.25 we can calculate the along-basin tidal currents across a section of arbitrary depth distribution H. The lateral tidal current may be obtained from

$$\hat{v}_0 = \frac{fg}{\omega^2}\partial\frac{\hat{\eta}_0}{\partial x}\left[1 - \frac{\cos h(\alpha_z z)}{\cos h(\alpha_z H)} + \frac{1}{2}\left((\alpha_z z)^2 - (\alpha_z H)^2\right)\right] + \frac{g}{2i\omega}\frac{\partial\hat{\eta}_0}{\partial y}\left[(\alpha_z z)^2 - (\alpha_z H)^2\right], \tag{3.26}$$

where the transverse water-level slope is:

$$\frac{\partial\hat{\eta}_0}{\partial y} = \frac{if}{\omega}\frac{\partial\hat{\eta}_0}{\partial x}\left[3\frac{\alpha_z H - \tan h(\alpha_z H)}{(\alpha H)^3} - 1\right]. \tag{3.27}$$

Solution 3.24, with 3.25, has two free parameters, A_z and $\overline{U_0}$, and describes the cross-channel distribution of the along-basin tidal current amplitude (e.g., Figure 3.8).

Although solutions 3.24 through 3.27 include complex variables, they can be represented in relatively straightforward programing, as in Animation 3.7. Tidal current amplitudes and phases derived with this solution (Figure 3.8) show clear bathymetric influences. Three main features are portrayed in these solutions. First, the strongest tidal currents are expected in the deepest part of the cross-section and at the surface, farthest from bottom frictional effects. Second, the phase of the tidal currents also follows the bathymetry closely. In the representation of Figure 3.8, the smallest phase values indicate the greatest delays in the response to tidal forcing; that is, tidal currents change earliest in Figure 3.8 where phase values are

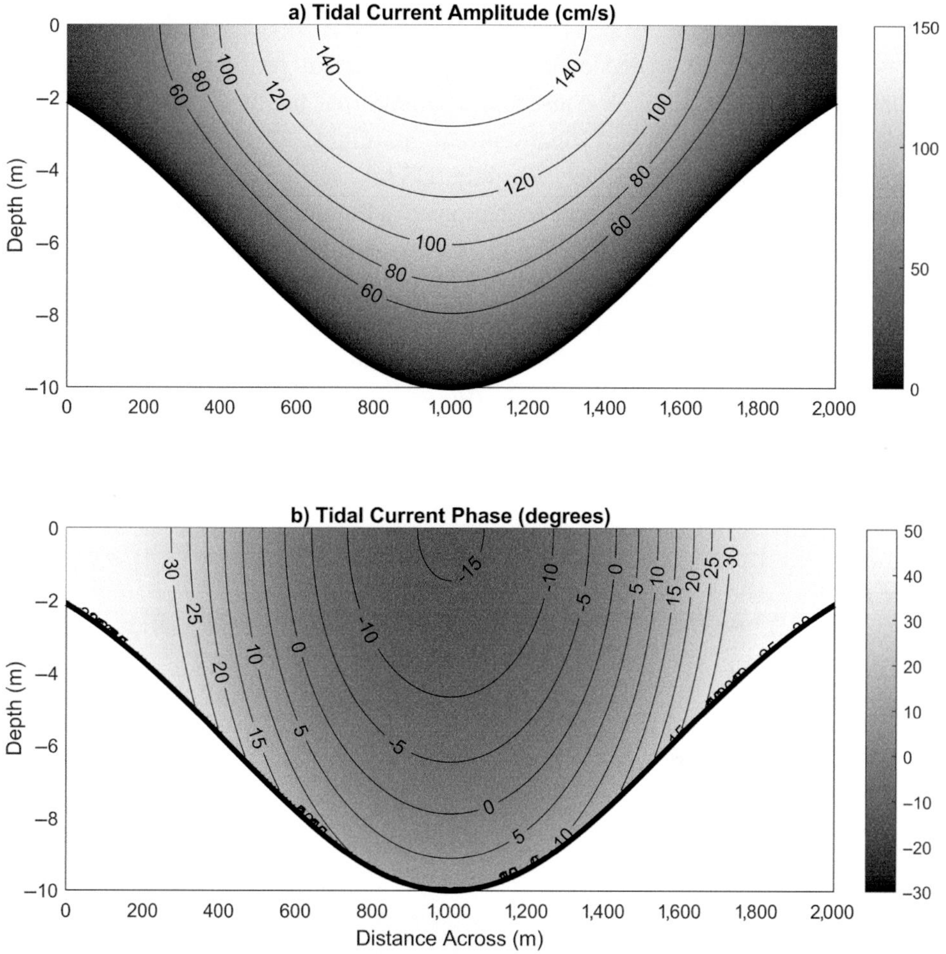

Figure 3.8 Tidal current amplitude and phase obtained from drawing equation 3.24 for an arbitrary depth $H(y)$ (see Animation 3.7). Maximum tidal currents occur at the surface, away from the effects of bottom friction. The greatest phase delays occur where the currents are strongest. The phase values in degrees can be converted to time (in minutes or hours) considering that one cycle (360°) takes 12 hours to complete. Following this reasoning, a phase lag of 1° is equivalent to a delay of 2 minutes.

greatest. Thus, tidal currents switch from flood to ebb, or from ebb to flood, earlier over the shoals of the cross-section relative to the deep channel. The third feature of the solution is that at any local velocity profile, tidal current switches earlier near the bottom compared to the surface. This phase lag is related to the fact that faster flows take longer to respond to a change in pressure gradient (from the tidal reversals) than slower flows. There are many observational examples that support these theoretical concepts. The next subsection illustrates one of the examples.

3.4.6 Observations Compared to Theoretical Solutions

Measurements with a towed acoustic Doppler current profiler over two cross-estuary transects and through consecutive semidiurnal tidal cycles have provided cross-section distributions of tidal current amplitude (Figure 3.9) and phase (Figure 3.10) in the James River, Virginia, USA.

Both measurements and model (equations 3.24 and 3.25, "Analytical Model") show maximum tidal current amplitude near the surface and over the deepest part of the transect. Also, the contours of tidal current amplitude nearly follow the bathymetric profile, an indication of bottom friction influence. In turn, the distribution of the phase lags in the transect shows similarities between the observations and the analytical results (Figure 3.10). Phase changes in the tidal currents occur first over shoals, compared to channel and near the bottom

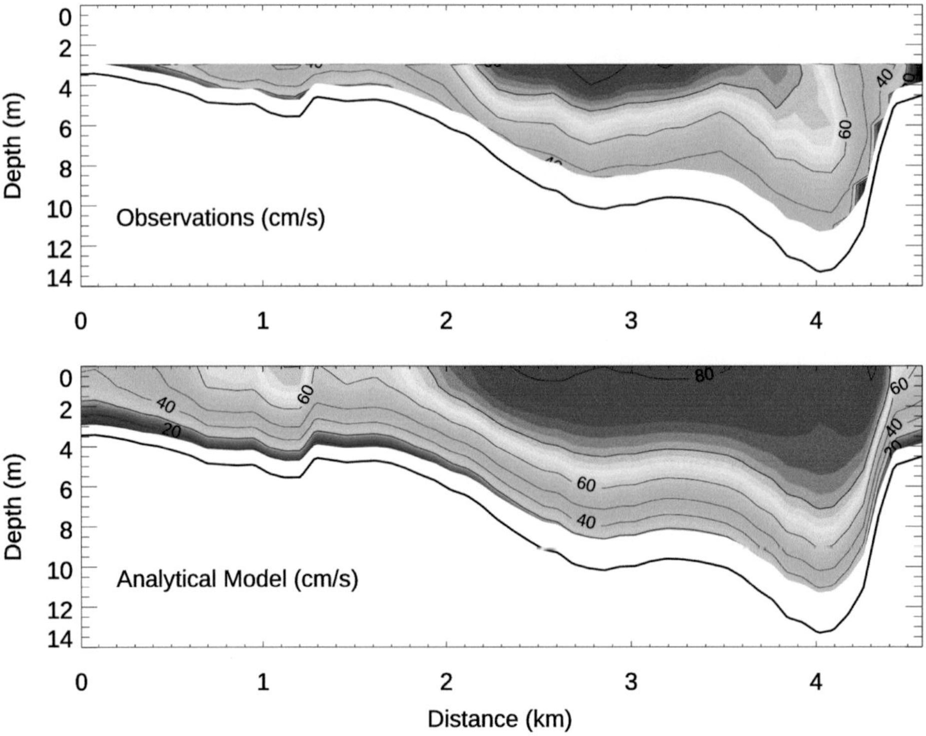

Figure 3.9 Comparison between semidiurnal tidal current amplitudes observed and modeled with equation 3.24 in the James River, Virginia, USA. Isolines nearly follow the bottom profile (black line). Masked portions above the bottom indicate portions of the water column unresolved by the current profiler. Maximum tidal currents appear in the portion of the cross-section with the lowest bottom frictional effects.

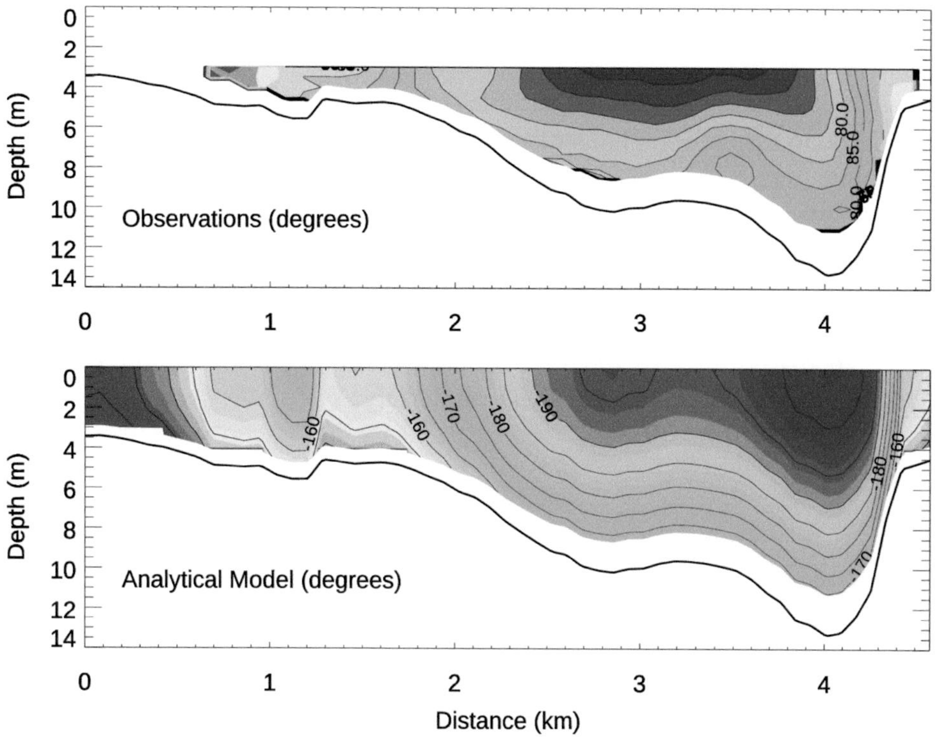

Figure 3.10 Comparison between tidal current phases observed and modeled with equation 3.24 in the James River, Virginia, USA. There is nearly a 40-minute delay between the tidal currents in the channel, relative to shoals, and close to 20-minute delay of currents at the surface compared to the bottom.

compared to the surface. This analytical approach has been used successfully to explain the tidal current distributions in cross-channel transects throughout the world. Differences between observations and model results can be attributed to processes neglected in the dynamics described in equations 3.23. Some of those processes ignored are advective accelerations (e.g., curvature and along-basin variability), baroclinicity (density gradients affecting tidal currents), and variable eddy viscosity in time and space.

It is evident that the spatial and temporal heterogeneity of tidal currents in semienclosed basins presents a challenge to sample-dissolved and suspended matter appropriately. It is expected that what happens biogeochemically in a channel will be different from what happens over shoals, and at the basin's head compared to its entrance. Even within any given profile, it is anticipated that there will be differences from surface to bottom. In addition, temporal variations in tidal currents will provide different situations during different tidal phases at any given location.

3.5 Take-Home Message

Tides are the most predictable phenomenon in the study of hydrodynamics in semienclosed basins. Therefore, an initial step in any hydrodynamic exploration should be to assess basic properties of tides: form factor, tidal range, dominant harmonics, and basin synchronicity. Furthermore, a distinction should be made regarding the relative importance between synodic and declinational fortnight and the implications of such influence. Also, the relevance of semiannual and annual harmonics should be established. Dynamically, tides in semienclosed basins can be studied with relatively simple approaches that consider frictionless scenarios. These approaches may be modified to allow the influence of friction, coastline convergence, the Earth's rotation, and bathymetry.

Additional Sources

Ensing, E., H.E. deSwart, and H. M. Schuttelaars (2015) Sensitivity of tidal motion in well-mixed estuaries to cross-sectional shape, deepening, and sea level rise. *Ocean Dyn.* 65: 933–950.

Huijts, K.M.H., H.M. Schuttelaars, H.E. de Swart, and A. Valle-Levinson (2006) Lateral entrapment of sediment in tidal estuaries: An idealized model study. *J. Geophys. Res. Oceans* 111, C12016, doi:10.1029/2006JC003615.

Officer, C.B. (1978) *Physical Oceanography of Estuaries (and Associated Coastal Waters).* New York: John Wiley and Sons, Inc.

Parker, B.B. (2007) Tidal analysis and prediction. Silver Spring, MD, NOAA NOS Center for Operational Oceanographic Products and Services, 378pp (NOAA Special Publication NOS CO-OPS 3). DOI: http://dx.doi.org/10.25607/OBP-191.

Ross, L., H. de Swart, E. Ensing, and A. Valle-Levinson (2017) Three-dimensional tidal flow in fjord-like basin with converging width: An analytical model. *J. Geophys. Res. Oceans* 122: 558–576.

4

Shallow-Water Tides

This chapter describes the mechanisms by which tides may be *deformed* or *distorted* or *rectified* (all terms referring to the same phenomenon) when they enter semienclosed basins. The chapter first presents qualitative and quantitative arguments for the presence of distortions and the generation of high-frequency harmonics, that is, of periodicities in the tide that are shorter than the fundamental (semidiurnal and diurnal) tidal periods. The chapter continues by exploring the possibilities of generation of high-frequency harmonics in what are referred to as *shallow-water tides*. These shallow-water tides can be *overtides* or *compound tides*, depending on what harmonics generate them. The chapter goes further with an explanation of the physical meaning of processes causing distortions. It concludes by providing a pair of examples of overtides and compound tides.

4.1 Tidal Distortions and the Appearance of High-Frequency Harmonics

Qualitatively, tidal *distortions* (or *rectifications* or *asymmetries*) may be identified by asymmetries in the shape of the tide as we observe its changes over time (Figure 4.1). Such tidal asymmetries may appear during high waters or low waters, or during flood or ebb currents. A record at the entrance to a semienclosed basin (e.g., Bar Harbor in Figure 4.1) will rarely show distortions. Further into the basin, a distorted tide may display, among various possibilities, flattened low-water periods (e.g., Conimicut Light, Figure 4.1), double-low waters (e.g., Providence, Figure 4.1), or shortened periods of low waters with relatively rapid (<5 h) rise from low to high tide (e.g., Newbold, Figure 4.1). Except for Bar Harbor, in Figure 4.1, all records indicate non-negligible influence from shallow-water tides, which manifest themselves as distortions to the tidal signal.

In any given semienclosed basin, the tide may become increasingly distorted in its progression along the basin, either as it amplifies or as it attenuates (e.g., Figure 4.2). In both examples of Figure 4.2, the rising phase of the tide becomes

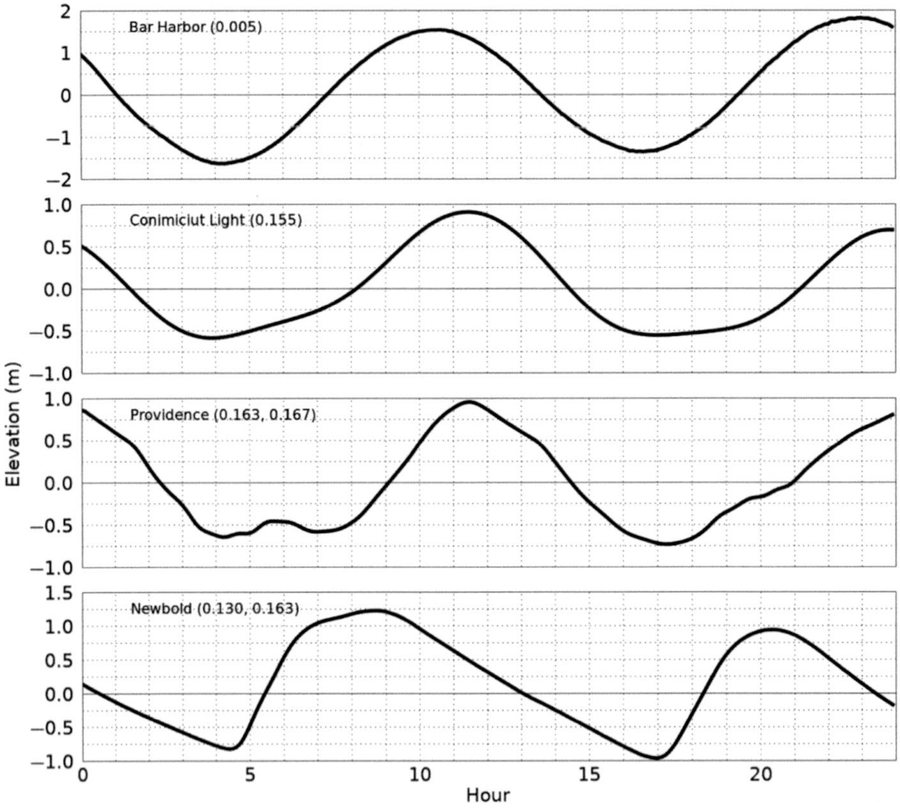

Figure 4.1 Tide records during one day from different locations on the east coast of the United States. Records illustrate distinct types of *asymmetries* (*distortions* or *deformations*) to the tidal wave. At Bar Harbor, within Frenchman Bay in Maine, there is no evidence for asymmetries. At Conimicut Light, in Narragansett Bay, Rhode Island, low-tide periods are flatter than high tide. At Providence, also in Narragansett Bay, the low tide may show "double-low" waters. At Newbold, Pennsylvania, in the upper tidal reaches of Delaware Bay, it takes the tide 3–4 hours to rise while it takes 8–9 hours to fall. Numbers in parentheses portray M_4/M_2 ratios, that is, measures of nonlinearities. For Providence and Newbold, the second number displays $[M_4+ M_6]/[M_2 + N_2]$ ratios, another measure of nonlinearities.

steeper than the falling phase as we move into the basin. Evidently, the tidal wave displays a nearly sinusoidal shape at the entrance and becomes highly deformed at the upstream reaches of the tidal signal. In the case that illustrates amplification (Figure 4.2, Delaware Bay), that is, *hypersynchronic* conditions, convergent coastlines overwhelm frictional effects (see Section 3.4.3). In the attenuated tidal wave of Figure 4.2 (Columbia River), *hyposynchronic* conditions result from

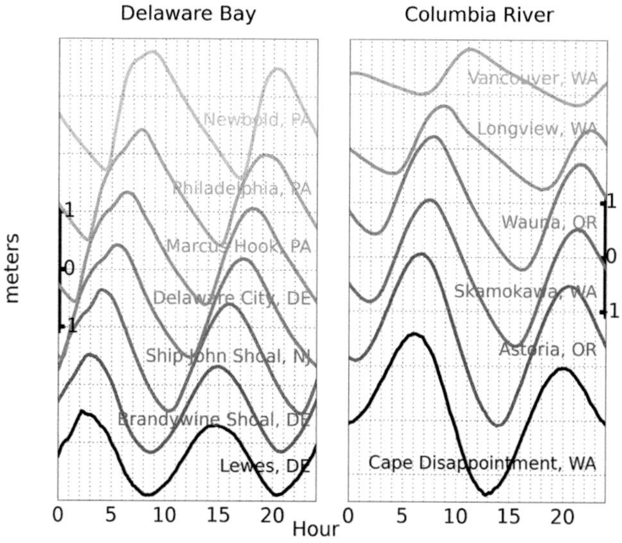

Figure 4.2 Tide records along Delaware Bay and the Columbia River estuaries. The records illustrate the phase propagation, the amplification or attenuation, and the distortion of the tidal signal. The time lag in high tides from entrance to the head is different from the low tide. In Delaware Bay, the high-tide lag between Lewes and Newbold is ~6.5 h, whereas the low-tide lag is ~8.5 h. In the Columbia River, the high-tide lag from Disappointment to Vancouver is ~5 h while the low-tide difference is ~9 h.

friction dominating coastline convergence. In general, deformations to the tide can be caused by (i) converging coastlines, (ii) river discharge competing against the tidal wave, (iii) difference in frictional effects between flood and ebb, (iv) spatial gradients in tidal currents, and (v) storm-driven currents that may compete with the tidal currents.

In the extreme tidal distortion along a semienclosed basin, rising water may develop over relatively short periods of only <2 h. Such a rapid rise can result in tidal bores like the one illustrated in Video 4.1. Tidal bores build during early flood tidal stages throughout more than 40 places in the world, mainly in basins with converging coastlines. The most spectacular tidal bores occur during the largest spring tides of the year and during relatively low river discharge conditions. Still, the influence of river discharge seems to be a key factor in their development. There are examples, such as the Colorado River delta in the upper Gulf of California, where there used to be tidal bores but which are now absent after reductions of river discharge to practically zero at the delta.

Quantitatively, the basic indicator that the tide is prone to distortion can be drawn from the concept of *conservation of mass* or *equation of continuity*. We can think of the deformation or distortion of a wave as it travels through a funnel-shaped vessel. Stretching the analogy further, think of your voice getting distorted when you speak through an elongated conic duct with a drastic, say 10 times, reduction in cross-sectional area. Similar to equation 3.4, but also considering changes of width along a basin (*B*(*x*)), that is, converging coastlines, the continuity equation may be written as

$$\frac{\partial \eta}{\partial t} + \frac{1}{B}\frac{\partial}{\partial x}[B(H+\eta)\,U] = 0, \tag{4.1}$$

where the mean water depth is H and the total water depth is $H + \eta$, which changes in time with the tide at any cross-section of the basin. Equation 4.1 describes changes in water elevation over time being linked to variations in width B, depth H, and depth-averaged tidal currents U along the basin. In the same equation, we can approximate both η and U as a sum of i number of harmonics, that is,

$$\begin{aligned} \eta &= \sum_i A_i \cos(\omega_i t) \\ U &= \sum_i U_{0i} \cos(\omega_i t) \end{aligned}. \tag{4.2}$$

For one harmonic, say $i = \mathrm{M_2}$, conservation of mass (equation 4.1) will be proportional to the product:

$$\eta U \sim \cos(\omega_{\mathrm{M_2}} t) \cdot \cos(\omega_{\mathrm{M_2}} t) = \cos^2(\omega_{\mathrm{M_2}} t). \tag{4.3}$$

But through a trigonometric identity:

$$\cos^2(\omega_{\mathrm{M_2}} t) = \frac{1}{2}(1 + \cos(2\omega_{\mathrm{M_2}} t)). \tag{4.4}$$

It follows, then, that in situations when conservation of mass is given by equation 4.1, without simplification (more on that below), a forcing (or mother) tidal harmonic, like the $\mathrm{M_2}$, will cause two additional harmonics. One is not really a harmonic because it has zero frequency (first term on the right-hand side of equation 4.4 that equals ½). The second harmonic attains twice the frequency of the mother harmonic. For the $\mathrm{M_2}$:

$$2\omega_{\mathrm{M_2}} = 2\underbrace{\left[\frac{2\pi}{12.42\ \mathrm{h}}\right]}_{\omega_{M_2}} = \frac{2\pi}{6.21\ \mathrm{h}} = \omega_{\mathrm{M_4}}. \tag{4.5}$$

The “child” harmonic of the $\mathrm{M_2}$ is the $\mathrm{M_4}$ and has one half its period (= 6.21 h), or twice its angular frequency. We call this child an “overtide.” An example of an $\mathrm{M_2}$ tide distorted by an $\mathrm{M_4}$ harmonic is shown in Animation 4.1, where individual harmonics are portrayed as lines and the sum of the two harmonics as the shaded wave.

A simple approach to determine the influence of nonlinear processes on tidal dynamics, that is, of tidal distortions, is to study the ratio M_4/M_2. The ratio examines the amplitude of tidal elevation or tidal current for the two harmonics. This ratio, however, can sometimes be untrustworthy. In Figure 4.1, the ratio for tidal amplitude is shown in parenthesis as the first or only number on each panel. The highly distorted tide at Newbold has a lower M_4/M_2 ratio (0.130) than at Providence (0.163). Nevertheless, inclusion of more overtides in the ratio, $[M_4 + M_6]/[M_2 + N_2]$, may provide further insights. On Figure 4.1 such a ratio is comparable between Providence and Newbold stations. In other instances, the ratio M_6/M_2 can also account for nonlinearities in the tidal signal. In the case of Figure 4.1, the ratio is 0.04 for Providence and 0.06 for Newbold.

The zero-frequency harmonic in equation (4.4), that is, ½, can also be referred to as a "tidal residual" after one or several cycles. We can say that tidal residuals derive from nonlinear interactions (product between harmonics) in the momentum and mass balance equations. Chapter 5 is dedicated to tidal residual flows.

Analogously, interaction of M_2 in η with S_2 in U, or the other way, yields

$$\eta U \sim \cos(\omega_{S_2} t) \cdot \cos(\omega_{M_2} t) = \underbrace{\cos(\omega_{M_2} + \omega_{S_2})\, t}_{\substack{\text{high freq distortion} \\ 6.1\ \text{h}}} + \underbrace{\cos(\omega_{M_2} - \omega_{S_2})\, t}_{\substack{\text{low freq modulation} \\ 14.8\ \text{d}}}, \tag{4.6}$$

again, using a trigonometric identity that indicates that the nonlinear interaction between M_2 and S_2, through conservation of mass, yields a *distortion* and a *modulation* of waves. The distortion is the sum of frequencies, resulting in a period of 6.1 h, a compound tide of M_2 and S_2. The modulation is the difference of frequencies, yielding a period of 14.8 d or a synodic fortnight. Strictly speaking, the *distortion* of any wave (surface, internal, acoustic, electromagnetic) arises when two frequencies are added and a *modulation* results when the frequencies subtract. We explore more on the distortions and modulations arising from interaction among different harmonics in the next section.

4.2 Interactions between Harmonics

Nonlinear terms (i.e., those involving products of oscillatory motions η and U) in equations 4.6 and 4.7 describe the two immediate consequences of the interaction between two harmonics: *distortion* and *modulation*. We can explore different

Table 4.1. *Interaction between the M_2 harmonic and the five harmonics on the first column**

	$\omega_{M2} - \omega_i$ (days)	$\omega_{M2} + \omega_i$ (hours)	$2\omega_{M2} - \omega_i$ (hours)	$2\omega_{M2} + \omega_i$ (hours)	$4\omega_{M2} - \omega_i$ (hours)
M_2 (12.42 h)	Residual (∞)	M_4 (6.21)	M_2 (12.42)	M_6 (4.14)	M_6 (4.14)
N_2 (12.66 h)	MN (27.5)	MN_4 (6.27)	$2MN_2$ (12.19)	$2MN_6$ (4.17)	$4MN_6$ (4.11)
S_2 (12.00 h)	MS (14.8)	MS_4 (6.10)	$2MS_2$ (12.87)	$2MS_6$ (4.09)	$4MS_6$ (4.19)
K_1 (23.93 h)	MK_1 (1.07)	MK_3 (8.17)	$2MK_3$ (8.38)	$2MK_5$ (4.93)	$4MK_7$ (3.57)
O_1 (25.82 h)	MO_1 (0.99)	MO_3 (8.38)	$2MO_3$ (8.17)	$2MO_5$ (5.00)	$4MO_7$ (3.52)

* Periods corresponding to each harmonic appear in parenthesis and are approximations in some cases. Actual periods are given by all decimals in Table 3.1 for each harmonic. Columns show periods in hours except for the second column of the table, which displays modulation periods in days. In addition to the second column, modulation periodicities are presented in the fourth and sixth columns. The third and fifth columns contain distortion periods.

possible distortions and modulations evolving from the interaction between M_2 and itself plus the other four principal harmonic constituents S_2, N_2, K_1, and O_1. These interactions are synthesized in Table 4.1, which describes the main *overtides* and *compound tides*.

Overtides and compound tides of M_2 tides with other semidiurnal harmonics are described along the top three rows of Table 4.1, underneath the header. The most frequently observed overtides and compound tides in regions dominated by semidiurnal tides are the quarter-diurnal harmonic, around a 6-h period, and the sixth-diurnal harmonic, around a 4-h period. All other combinations, on the fourth and fifth rows of the table, are compound tides between semidiurnal and diurnal periods. Under the dominance of mixed tides, the most frequently observed compound tide is the terdiurnal, with periods of around 8 h. The subindex of each harmonic constituent in the table indicates the number of cycles per period (e.g., day) of 24–25 h. The MN and MS modulations of semidiurnal tides correspond to the anomalistic monthly harmonic M_m (= 27.5 d) and to the synodic fortnightly harmonic MS_f (= 14.8 d).

Overtides of the K_1 harmonic (not shown in Table 4.1) are 23.93/2 = 11.96 h and 23.93/3 = 7.98 h, and of the O1 harmonic are 12.91 and 8.61 h. Curiously, these diurnal overtides are around the semidiurnal band (12 h) and near the band of interaction between semidiurnal and diurnal constituents (8 h). Therefore, distortions to the diurnal harmonics will reflect in reinforcement of semidiurnal tides and compound tides at the period of around 8 h.

4.3 Physical Explanation for Distortions

In addition to the M_4/M_2 or M_6/M_2 ratios, conservation of mass (equation 4.1) can provide another initial diagnostic for tidal distortions and modulations arising from

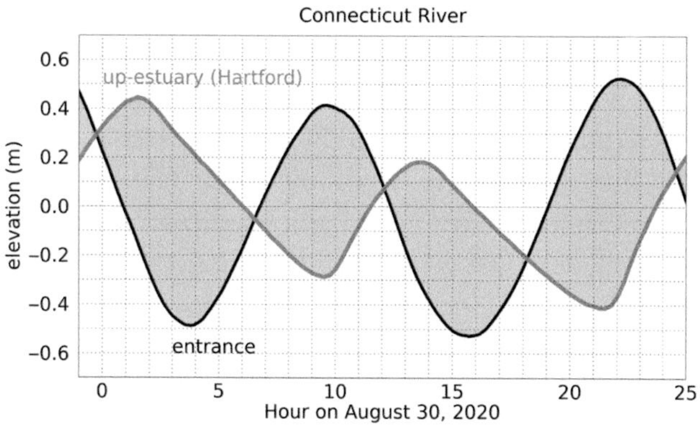

Figure 4.3 Tidal records for two semidiurnal cycles at the entrance to the Connecticut River and further up-estuary at Hartford, Connecticut. Note the signal attenuation, its distortion, and the asymmetric distribution of shaded areas in each half-cycle. This happens four times a day, consistent with the quarterdiurnal harmonic.

nonlinear interactions between harmonics. Such a diagnostic is made by comparing the relative size between η and H, that is, the ratio between the tidal amplitude A (e.g., equation 3.16) and the local water depth H. In general, when the ratio $A/H > 0.1$, that is, when the tidal amplitude is a non-negligible portion of the water depth, the tide is expected to show asymmetries. When $A \ll H$, equation 4.1 becomes equation 3.4, which means that the tidal wave behaves linearly, as the waves described in Chapter 3. Those linear tidal waves propagate at a phase speed of $\sqrt{gH}$. Instead, when $A/H > 0.1$, the wave dynamics are affected by nonlinear processes, that is, by the product ηU in equation 4.1. When $A/H > 0.1$, the tide propagates at an approximate phase speed of $\sqrt{g(H+\eta)}$ in such a way that the phase speed at high tide, the wave crest, is $\sqrt{g(H+A)}$. Similarly, the phase speed of the tidal wave at low tide, the wave trough, is $\sqrt{g(H-A)}$. The difference in phase speeds between high and low tides results in records like those observed in the upper estuary portions (e.g., Philadelphia, Newbold, Longview, and Vancouver) in Figure 4.2, where the tide is most distorted.

Yet another way of understanding the generation of overtides is by looking at the distortion of the tidal wave at up-estuary reaches compared to the entrance (Figure 4.3). Once more, we can see at the up-estuary record the relatively rapid rise of the tide relative to the span of its fall. The asymmetric distribution of shaded area per half-tidal cycle in the figure indicates the leakage of energy between the mouth and head of the system at a frequency of four times per day. This is a simplistic way of explaining the appearance of quarterdiurnal harmonics through nonlinear processes.

We say that the tidal wave is affected by nonlinear processes when it is distorted because the nonlinear terms in the mass (continuity) and momentum balances become non-negligible. We discussed the mass balance (equation 4.1) having a nonlinear contribution from the term $\frac{1}{B}\frac{\partial}{\partial x}[B\,\eta\,U]$, proportional to $\eta\,U$. The dynamic contributions from nonlinearities to the momentum balance can be gleaned from the balance for a tidal wave affected by friction (equation 3.14) and also considering tidal flow divergences along the basin (advective accelerations or advective fluxes of momentum per unit mass). Nonlinearities arise from advection and from the frictional term that considers tidal variations on the total water depth, that is, $H+\eta$:

$$\underbrace{\frac{\partial U}{\partial t}+U\frac{\partial U}{\partial x}}_{\textit{local \& advective accelerations}} = \underbrace{-g\frac{\partial \eta}{\partial x}}_{\textit{pressure gradient}} \underbrace{-\frac{C_b u_b|u_b|}{H+\eta}}_{\textit{bottom friction}}. \tag{4.7}$$

From this momentum balance we can identify two other mechanisms that distort tides into higher-frequency harmonics and modulate them to lower frequency. These are the advective momentum flux, as $U\frac{\partial U}{\partial x}$ is proportional to U^2, and bottom friction, as it is proportional to $u_b|u_b|$, where u_b is the near-bottom tidal current. Because equation 4.7 disregards transverse flows V, the magnitude of the near-bottom velocity is $|u_b|$.

Conceptually, the mechanism of tidal advective momentum flux can be explained from the rough relationship between tidal current amplitude U_0 and tidal amplitude A, e.g., equation 3.7, or

$$U_0=\frac{C}{H}A. \tag{4.8}$$

Typically, $C \gg U_0$, that is, the tidal phase speed, is much greater than the maximum tidal currents. But when $A/H > 0.1$, U_0 is not negligible anymore with respect to C. Consequently, the phase propagation speed of the tidal wave at high tide would become $C + U_0$, while at low tide it would be $C - U_0$. The result is an interaction of the tidal wave itself with the tidal current: faster switch from flood to ebb than from ebb to flood, as illustrated in Figure 4.4.

The mechanism that distorts tidal waves through bottom friction, proportional to $u_b|u_b|/(H+\eta)$ (see equation 4.7), is related to asymmetric loss of momentum flux per unit mass. Momentum losses caused by bottom friction will be greater during low water $(H-A)$ than at high water $(H+A)$ because $A/H > 0.1$. Moreover, the maximum momentum loss within each phase of the tidal cycle will appear at extreme currents, causing further distortions. The three mechanisms that favor nonlinear dynamic interactions between tidal harmonics and drive distortions and modulations to the tide are thus continuity, advective flux, and frictional losses.

Two other forcing agents can produce distortions to tidal waves. These agents are river discharge and wind-driven currents. They will be capable of modifying

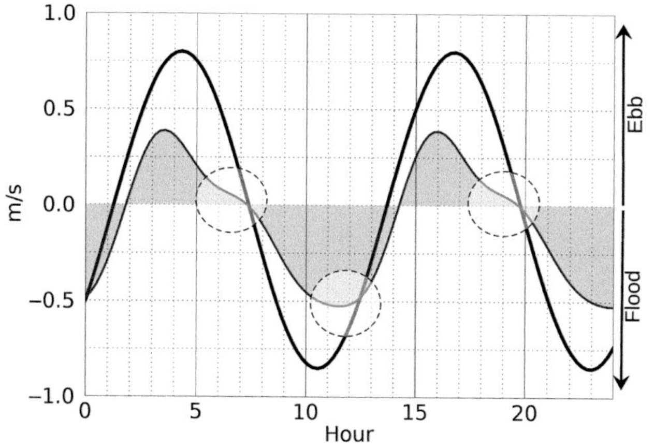

Figure 4.4 Synthetic tidal currents at the entrance to (thick black line) and in the interior of (shaded gray line) a semienclosed basin during two tidal cycles. The figure illustrates a sinusoidal flow at the entrance that is expected to retain the shape without distortion. The signal in the interior has additional contributions from M_4 and M_6. The flow goes through distortion in the interior. Ebb currents reach their peak and weaken toward slack waters earlier than expected (from the tidal behavior at the entrance) because of the interaction $C - U_0$. Similarly, flood currents remain in maximum flood and change to ebb later but over a shorter span than at the entrance because of the interaction $C + U_0$.

the tidal signal when the velocity of the river flow or of the wind-driven flow is comparable to the tidal currents. Increased river discharge, as during spring freshets and storm-induced river flooding, will tend to favor ebb currents and oppose flood currents. Such augmented asymmetry in tidal currents will redistribute the duration of nonlinear effects from bottom frictional losses. Increased momentum flux loss, relative to pre-river pulse conditions, and for longer spans, will occur during ebb rather than flood. Consequences will include attenuation of the tidal signal and generation of overtides and compound tides. An analogous response is expected when unidirectional and sustained (at least for one tidal cycle) wind-driven currents interact with tidal signals, as during hurricanes or tropical storms.

4.4 Examples

A practical illustration of tidal distortion through overtides presents data from the Ems river, Germany, essentially at the border with The Netherlands in the Wadden Sea. Data correspond to 102 repetitions of the same cross-estuary transect throughout one semidiurnal tidal cycle. Measurement points represent cross-sectional averages of along-estuary tidal currents (Figure 4.5). The tidal current variations during the cycle sampled are dominated by the semidiurnal harmonic (Figure 4.5a). Sampling lasted only one semidiurnal cycle, making it unrealistic to

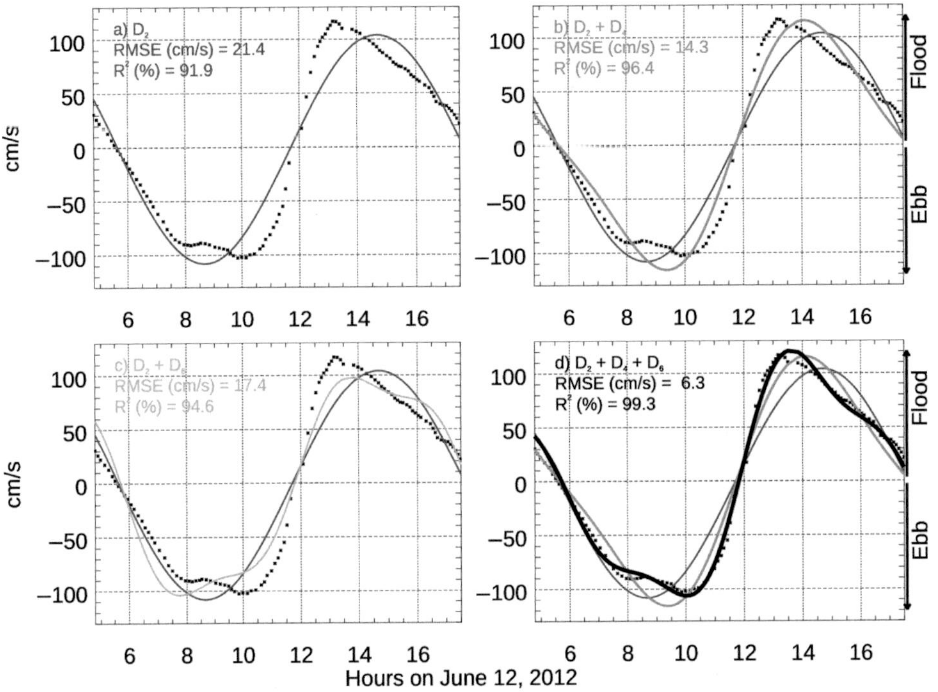

Figure 4.5 Time series of 102 cross-sectional mean flows along the Ems river (symbols) compared to different combinations of frequency bands (continuous lines). Each panel displays the band fitted to the time series, via least squares fit, following equation 4.9. Each panel also shows the root-mean-squared-error (RMSE) of the fit and its variance explained (R^2). In (a) the only continuous line is the semidiurnal harmonic, which is also drawn in the other three panels. In (b) a slightly thicker gray line portrays the fit with the semidiurnal and quarterdiurnal bands, together. In (c) the light gray line is the result of the semidiurnal and sixthdiurnal combined. Panel (d) exhibits the best fit (black, thick line) that includes the three bands: semidiurnal, quarterdiurnal, and sixthdiurnal.

resolve different semidiurnal harmonic constituents (M_2, S_2, N_2). We therefore refer to the semidiurnal band D_2 because at the time of the particular cycle sampled, all semidiurnal constituents were interacting but unresolved individually. The constituents interacted to produce the amplitude and phase of the semidiurnal band observed. The contribution from the semidiurnal band to the observed data points u_m can be drawn from an expression that approximates the observations in terms of a residual flow u_r plus the sum of i harmonics with amplitude u_{0i}, phase ϕ_i, and known frequency ω_i:

$$u_m = u_r + \sum_{i=1}^{M} u_{0i} \sin\left(\omega_i t + \phi_i\right). \tag{4.9}$$

For $M = 1$ harmonic, which will be the semidiurnal band, equation 4.9 can be written as

$$u_m = \boldsymbol{u_r} + \boldsymbol{u_{0D2}} \sin\left(\omega_{D2}t + \boldsymbol{\phi_{D2}}\right). \tag{4.10}$$

The parameters needed to match measured data points u_m with the right-hand side of equation 4.10, also referred to as free parameters, are $\boldsymbol{u_r}$, $\boldsymbol{u_{0D2}}$, and $\boldsymbol{\phi_{D2}}$, which appear in bold. These parameters are obtained from minimizing the squared error (least squares fit) between the two sides of equation 4.10. Details on the "least squares fit" approach are given in the appendix to this chapter (Appendix 4.1). For the observed data, we can see (Figure 4.5a, continuous black line) that the semidiurnal band D_2 accounts for close to 90% of the variance of the signal. The result is practically the same if the period of the semidiurnal band chosen for ω_{D2} is 12.42 h or simply 12 h. Although D_2 accounts for a great portion of the variance, we can see that the maxima and their timing do not match the observations. Discrepancies are related to the influence of overtides.

By adding the quarterdiurnal band D_4 to equation 4.9, instead of having three free parameters, we now have five:

$$u_m = \boldsymbol{u_r} + \boldsymbol{u_{0D2}} \sin\left(\omega_{D2}t + \boldsymbol{\phi_{D2}}\right) + \boldsymbol{u_{0D4}} \sin\left(\omega_{D4}t + \boldsymbol{\phi_{D4}}\right) \tag{4.11}$$

that is, the amplitude and phase for the semidiurnal and quarterdiurnal bands, and the residual flow. Minimizing the squared errors between both sides of equation 4.11 provides a better match than with only the semidiurnal band (Figure 4.5b, gray line). Still, the shape of the observed tidal peaks are flatter during ebb (negative values) and sharper during flood (positive values). The combination of semidiurnal and sixthdiurnal bands (Figure 4.5c, light gray line) yields reasonable results but not better than those with D_2 and D_4.

The match improves noticeably when we combine D_2, D_4, and D_6 (Figure 4.5d, black thickest line). The flat curve around peak ebb and the sharpness around peak flood are represented well with these three harmonics. Results in all four panels of Figure 4.5 display negligible changes if instead of using 12.42, 6.21, and 4.14 h as the periods of the three bands prescribed, we use 12, 6, and 4 h. Data in this figure are also provided for an exercise to extract harmonics from the signal.

Data from the Suwannee River on Florida Peninsula's west coast provide an example of compound tides, as well as overtides. The site features a *form factor* (equation 3.2) of 0.65, that is, mixed tides with semidiurnal dominance. The water elevation at this site (Figure 4.6) exhibits the mixed character with *synodic* (thick black line) and *declinational* (thick gray line) fortnightly modulation. It is evident that the maxima in synodic fortnightly modulation coincide with greatest semidiurnal tides and maxima in declinational fortnightly signal show the largest diurnal tides. Tidal currents in this example follow water elevation closely, with semidiurnal currents being more noticeable than semidiurnal water elevations

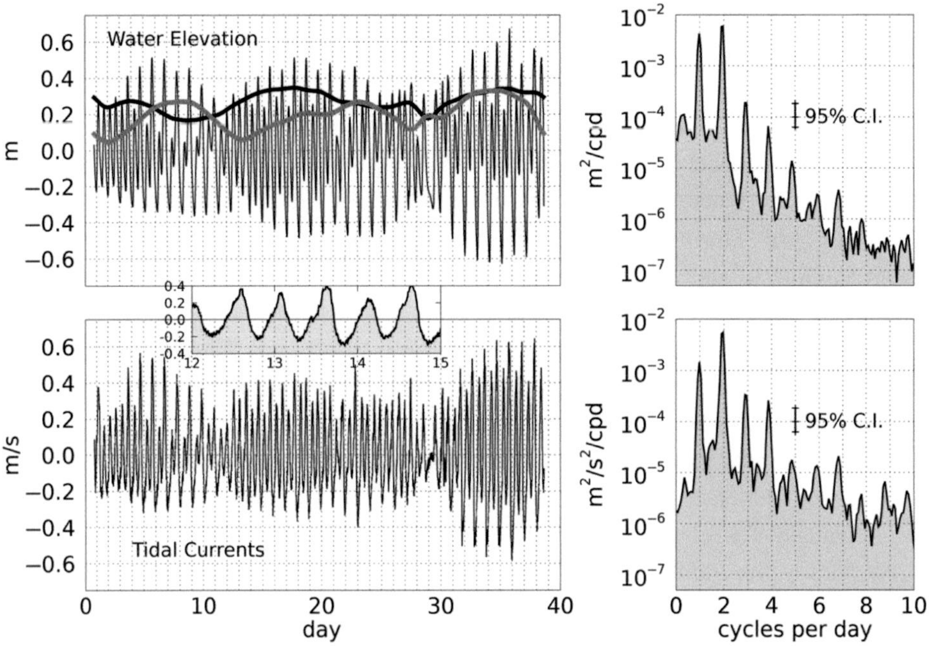

Figure 4.6 Tidal-elevation and current records from the Suwannee River, Florida, on the western coast of the Florida Peninsula. The water elevation time series also shows the amplitude of the semidiurnal band (thick black line) and of the diurnal band (thick gray line). For the currents, positive values indicate flooding into the estuary. Also shown is a three-day portion to illustrate tidal distortions. The right panels show the power spectra of their respective record to the left. The vertical axis is in logarithmic (base 10) scale. The confidence interval (95% C.I.) for each spectral estimate is illustrated by one bar at five cycles per day.

because of the flow response to changes in elevation over time. An enlarged view to three days of tidal currents (days 12, 13, and 14 on Figure 4.6) displays clear distortions with "peaky" floods (positive values) after a longer switch from maximum ebb to maximum flood than from maximum flood to maximum ebb.

Simple inspection of the three-day tidal current record from the Suwannee River suggests the appearance of shallow-water tides. Because it is a mixed tides basin, we should expect to see overtides of the semidiurnal constituents and compound tides between semidiurnal and diurnal harmonics (see Table 4.1). In fact, we can decompose or deconstruct both water-level and tidal-current records in terms of their frequency variance; that is, we can draw power spectral descriptions (or Fourier decompositions) of these signals (also shown in Figure 4.6). These representations reveal, as expected, that the variance of the water-level and tidal-current records are dominated by the semidiurnal and diurnal bands. Because of the temporal resolution of the records and the increased reliability of each variance estimate, we are unable to resolve specific constituents, only frequency bands. The frequency sorting in Figure 4.6 also unveils several shallow-water constituents. The most prominent are the terdiurnal and quarterdiurnal bands but there is also noticeable influence of the 5, 6, and 7 cycles per day frequency bands.

4.5 Take-Home Message

Tides will deform in semienclosed basins and display overtide and compound tide variability. These distortions are also known as shallow-water tides. Distortions are caused by (i) converging coastlines, (ii) river discharge competing against the tidal wave, (iii) difference in frictional effects between flood and ebb, (iv) spatial gradients in tidal currents, and (v) storm-driven currents that may compete with the tidal currents. These agents enter the momentum balance through nonlinear advection and bottom friction, and the mass balance through flow convergence.

Appendix 4.1: Further Information

This appendix presents a method to calculate tidal harmonics from a record of water level or tidal currents like that presented in Figure 4.5, using equations 4.9–4.11. The appendix may seem to have many equations. It uses mathematical concepts covered over the basic training of every scientist and engineer. So, with a little patience, the contents of this appendix should be decipherable and handy.

Taking equation 4.10, measurements are represented by u_m, and we are seeking to obtain the optimal values of $\boldsymbol{u_r}$, $\boldsymbol{u_{0D2}}$, and $\boldsymbol{\phi_{D2}}$ that validate equation 4.10, which can be rewritten with the trigonometric identity,

$$\sin(A+B) = \cos B \sin A + \cos A \sin B,$$

as

$$u_m = \underbrace{\boldsymbol{u_r} + \boldsymbol{u_1}\sin(\omega_{D2}t) + \boldsymbol{u_2}\cos(\omega_{D2}t)}_{u'}, \tag{A4.1}$$

by defining

$$\boldsymbol{u_1} = \boldsymbol{u_{0D2}}\cos\boldsymbol{\phi_{D2}}; \quad \text{and} \quad \boldsymbol{u_2} = \boldsymbol{u_{0D2}}\sin\boldsymbol{\phi_{D2}}. \tag{A4.2}$$

We are trying to minimize the difference between both sides of equation A4.1. The squared errors between measurements u_m and the harmonic representation u' may be expressed as ε^2:

$$\varepsilon^2 = \sum [u_m - u\prime]^2 = \sum u_m^2 - 2u_m u' + u\prime^2,$$

where the sum is over M number of measurements. Then, using A4.1 for u', we obtain the squared errors as (the bit challenging part, but not really, could be expanding $u\prime^2$):

$$
\begin{aligned}
\varepsilon^2 = \sum\{&u_m^2 - 2u_m u_r - 2u_m u_1 \sin(\omega_{D2}t) \\
&- 2u_m u_2 \cos(\omega_{D2}t) + u_r^2 + 2u_r u_1 \sin(\omega_{D2}t) \\
&+ 2u_r u_2 \cos(\omega_{D2}t) \\
&+ 2u_1 u_2 \sin(\omega_{D2}t) \cos(\omega_{D2}t) \\
&+ u_1^2 \sin^2(\omega_{D2}t) + u_2^2 \cos^2(\omega_{D2}t)\}
\end{aligned}
$$

Then, to find the minimum distance between measured u_m and theoretical u' values we need to minimize ε^2 with respect to the three free parameters u_r, u_1, and u_2, that is, make the differentials $\partial\varepsilon^2/\partial u_r$, $\partial\varepsilon^2/\partial u_1$, and $\partial\varepsilon^2/\partial u_2$ equal to zero (from differential calculus, a minimum is found by making the derivative or differential equal to zero).

Differentiating the error squared with respect to the first parameter (might seem ominous, but we see that only a few terms have u_r, so it should be straightforward):

$$
\partial\varepsilon^2/\partial u_r = \sum\{-2u_m + 2u_r + 2u_1 \sin(\omega_{D2}t) + 2u_2 \cos(\omega_{D2}t)\} = 0; \tag{A4.3}
$$

then with respect to u_1 (not tough):

$$
\begin{aligned}
\partial\varepsilon^2/\partial u_1 = \sum\{&-2u_m \sin(\omega_{D2}t) + 2u_r \sin(\omega_{D2}t) \\
&+ 2u_2 \sin(\omega_{D2}t)\cos(\omega_{D2}t) + 2u_1 \sin^2(\omega_{D2}t)\} = 0;
\end{aligned} \tag{A4.4}
$$

and finally, with respect to u_2 (not bad by now):

$$
\begin{aligned}
\partial\varepsilon^2/\partial u_2 = \sum\{&-2u_m \cos(\omega_{D2}t) + 2u_r \cos(\omega_{D2}t) \\
&+ 2u_1 \sin(\omega_{D2}t)\cos(\omega_{D2}t) + 2u_2 \cos^2(\omega_{D2}t)\} = 0.
\end{aligned} \tag{A4.5}
$$

Remember that we are doing all this to find the optimal parameters, bold in equation A4.1, to match observations with the model described by such an equation. We can rearrange equations A4.3 through A4.5 by isolating the measurements and eliminating the 2s, which appear on every term:

$$
\begin{aligned}
&\sum\{u_m = u_r + u_1 \sin(\omega_{D2}t) + u_2 \cos(\omega_{D2}t)\} \\
&\sum\{u_m \sin(\omega_{D2}t) = u_r \sin(\omega_{D2}t) + u_2 \sin(\omega_{D2}t)\cos(\omega_{D2}t) + u_1 \sin^2(\omega_{D2}t)\} \\
&\sum\{u_m \cos(\omega_{D2}t) = u_r \cos(\omega_{D2}t) + u_1 \sin(\omega_{D2}t)\cos(\omega_{D2}t) + u_2 \cos^2(\omega_{D2}t)\}.
\end{aligned}
$$

This system of 3 equations can be written in matrix form:

$$\underbrace{\begin{bmatrix} \sum u_m \\ \sum u_m \sin(\omega_{D2}t) \\ \sum u_m \cos(\omega_{D2}t) \end{bmatrix}}_{\substack{\textit{Measurements} \\ \mathbf{B}}}$$

$$= \underbrace{\begin{bmatrix} M & \sum \sin(\omega_{D2}t) & \sum \cos(\omega_{D2}t) \\ \sum \sin(\omega_{D2}t) & \sum \sin^2(\omega_{D2}t) & \sum \sin(\omega_{D2}t)\cos(\omega_{D2}t) \\ \sum \cos(\omega_{D2}t) & \sum \sin(\omega_{D2}t)\cos(\omega_{D2}t) & \sum \cos^2(\omega_{D2}t) \end{bmatrix}}_{\substack{\textit{coefficients} \\ \mathbf{A}}} \underbrace{\begin{bmatrix} u_r \\ u_1 \\ u_2 \end{bmatrix}}_{\substack{\textit{unknowns} \\ \mathbf{X}}}.$$

If **B** is the matrix of measurements, **A** the matrix of coefficients, and **X** the unknowns, then the system of equations to solve is $\mathbf{B} = \mathbf{AX}$. The solution of the system of equations is $\mathbf{X} = \mathbf{A}^{-1}\,\mathbf{B}$. Matrix **B** is built with the measurements u_m collected at time t and the harmonic ω_{D2}. Matrix **A** is symmetric and is built with the harmonic ω_{D2} and the time t of measurements, where M is the number of measurements to go in the fit. All sums in the matrices are over M number of measurements. The solution **X** requires a matrix inversion of **A** with subsequent multiplication times **B** (see Matlab exercise).

But matrix **X** is not the final result. In that matrix, u_r represents the residual flow, while the amplitude of the harmonic ω_{D2} is

$$u_{0D2} = \sqrt{u_1^2 + u_2^2}, \tag{A4.6}$$

and its phase is

$$\phi_{D2} = \tan^{-1}(u_2/u_1). \tag{A4.7}$$

This gives us the three free parameters sought u_r, u_{0D2}, and ϕ_{D2}. These parameters are used in equation 4.10 to approximate the measurements u_m with one harmonic.

For two harmonics ω_{D2} and ω_{D4}, as in equation 4.11, matrix **A** becomes

$$\underbrace{\begin{bmatrix} M & \sum\sin(\omega_{D2}t) & \sum\cos(\omega_{D2}t) & \sum\sin(\omega_{D4}t) & \sum\cos(\omega_{D4}t) \\ \sum\sin(\omega_{D2}t) & \sum\sin^2(\omega_{D2}t) & \sum\sin(\omega_{D2}t)\cos(\omega_{D2}t) & \sum\sin(\omega_{D2}t)\sin(\omega_{D4}t) & \sum\sin(\omega_{D2}t)\cos(\omega_{D4}t) \\ \sum\cos(\omega_{D2}t) & \sum\sin(\omega_{D2}t)\cos(\omega_{D2}t) & \sum\cos^2(\omega_{D2}t) & \sum\cos(\omega_{D2}t)\sin(\omega_{D4}t) & \sum\cos(\omega_{D2}t)\cos(\omega_{D4}t) \\ \sum\sin(\omega_{D4}t) & \sum\sin(\omega_{D2}t)\sin(\omega_{D4}t) & \sum\cos(\omega_{D2}t)\sin(\omega_{D4}t) & \sum\sin^2(\omega_{D4}t) & \sum\sin(\omega_{D4}t)\cos(\omega_{D4}t) \\ \sum\cos(\omega_{D4}t) & \sum\sin(\omega_{D2}t)\cos(\omega_{D4}t) & \sum\cos(\omega_{D2}t)\cos(\omega_{D4}t) & \sum\sin(\omega_{D4}t)\cos(\omega_{D4}t) & \sum\cos^2(\omega_{D4}t) \end{bmatrix}}_{\substack{\textit{coefficients} \\ \mathbf{A}}}$$

whereas **B** and **X** become

$$\mathbf{B} = \begin{bmatrix} \sum u_m \\ \sum u_m \sin(\omega_{D2}t) \\ \sum u_m \cos(\omega_{D2}t) \\ \sum u_m \sin(\omega_{D4}t) \\ \sum u_m \cos(\omega_{D4}t) \end{bmatrix}, \quad \text{and} \quad \mathbf{X} = \begin{bmatrix} u_r \\ u_1 \\ u_2 \\ u_3 \\ u_4 \end{bmatrix}.$$

The amplitude and phase of the first harmonic are determined with equations A4.6 and A4.7. The same relationships apply for the second harmonic, but with u_3 and u_4 instead of u_1 and u_2, respectively.

We can use different metrics for the quality of the match or fit between u' and u_m, three of which are provided, goodness of fit (R^2), skill (S), and root–mean–square error (RMSE):

$$R^2 = \frac{\sum [\bar{u}_m - u']^2}{\sum [\bar{u}_m - u_m]^2}, \quad S = 1 - \frac{\sum [|u_m - u'|]^2}{\sum [|u' - \bar{u}_m| + |u_m - \bar{u}_m|]^2}, \quad \text{and}$$

$$\text{RMSE} = \sqrt{\frac{1}{M}\sum [u_m - u']^2}.$$

In all these metrics, the sums are over all M number of observations. The first two, R^2 and S, are dimensionless and the closer they are to 1, the better the modeled representation. The RMSE has the same units as the measurements and ideally should be close to zero or a small percentage (<10%) of the amplitude of the variations being modeled or predicted. In this case, the variations being modeled are approximated with the harmonics representation in equation 4.9.

Additional Sources

Parker, B.B. (1991) Nonlinear tidal interactions in shallow water. In *Tidal Hydrodynamics*. Edited by B.B. Parker. Hoboken, NJ: Wiley and Sons.

Parker, B.B. (2007) Tidal analysis and prediction. Silver Spring, MD, NOAA NOS Center for Operational Oceanographic Products and Services, 378 pp (NOAA Special Publication NOS CO-OPS 3). DOI: http://dx.doi.org/10.25607/OBP-191.

5

Tidal Residual Flows in Homogeneous, Semienclosed Basins

This chapter describes tidal residual flows (remaining after one or more tidal cycles) expected from dynamic considerations in semienclosed basins without the influence of density gradients. Keep in mind, as in other chapters, that this is intended as an introductory text. The material in this chapter could delve into advanced concepts but it stops at, perhaps, an intermediate level. To dive deeper into this topic, the reader could consult the additional readings suggested at the end. It is crucial to concentrate here on the physical processes and to consider the mathematical expressions as "shorthand" representations of those physical processes.

The chapter first qualitatively connects the concepts of Chapter 4, shallow-water tides, to the generation of residual flows. It then introduces a qualitative and quantitative description of the mechanisms that may generate these tidal residual flows in basins with a main channel flanked by shoals. It follows with an explanation of the flow structure in this type of basin, considering "long" and "short" basins, relative to the tidal wavelength. It then depicts examples that compare observations to theoretical expectations. The chapter concludes with descriptions of flows around headlands or sites with coastline curvature.

5.1 Residual Flows from Asymmetries in Tides under Homogeneous Fluids

As described in Chapters 3 and 4, purely sinusoidal tidal flows are symmetric during rising and falling tides and the average after one or several cycles results in zero residual flows. However, distortions to tidal waves characterize asymmetries that involve compound tides and overtides, but also result in tidally averaged residual flows directed either into or out of a semienclosed basin. Also, as stated in Chapter 4, such asymmetries can arise from mass conservation, from the advective flux of momentum, or through differential bottom friction effects. All of these mechanisms involve nonlinear interactions in shallow waters between flow and water elevation, or among flows at different locations in the basin. These mechanisms produce higher-frequency harmonics than the mother or forcing harmonics (e.g., M_2 and S_2) through the sum of their frequencies. But they can also generate zero-frequency flows or residual flows that may be modulated at relatively low frequencies equivalent to the difference between their frequencies. This modulation, caused by nonlinear interactions among harmonics (Table 4.1), should enhance the tidal residual flows that may arise from astronomic forcing at the same period of the modulations (e.g., synodic fortnight (M_2 and S_2) and anomalistic month (M_2 and N_2), see Section 3.3).

Using the concept that tidal residual flows may be modulated by the nonlinear interaction between two harmonics, residual flows during spring tides are expected to be larger than at neaps. Clearly, this would apply at locations primarily forced by tides where the residual flows are dominated by tidal deformations or rectifications (over wind or density gradients). Some examples of such locations are coastal lagoons during periods of wind relaxation or oscillatory homogeneous flows around a bend or a headland, through enhancement of the mechanisms responsible for tidal residual flows.

Qualitatively, from conservation of mass we can see that in an asymmetric wave there could be more water entering a semienclosed basin during flood than during ebb. This is analogous to the extreme case in which a breaking wave at the surf zone transports mass toward the beach. The wave-induced transport occurs in the direction of wave propagation and is known as *Stokes transport* (Figure 5.1a). In semienclosed basins, then, asymmetric tides should produce a net transport toward the basin's head (Figure 5.1a). This net transport will drive a water level set-up toward the head of the basin (Figure 5.1b), which will in turn force a return flow out of the basin. The return flow would be analogous to an undertow at the beach. In two dimensions (along-basin x and vertical z), the overall mass of water in the basin would be balanced between the Stokes transport into the basin and the return transport associated with the water-level slope. A residual flow profile will depend on the vertical distribution of the Stokes transport (per unit depth) and the return flow, as established by frictional effects. These two processes assume lateral homogeneity in bathymetry. They produce residual (or low-frequency) currents analogously to the conservation of mass mechanism described for overtides in Chapter 4.

Moreover, considering the lateral direction (y) and laterally varying bathymetry composed of a deep channel flanked by shoals, tidal currents will be strongest in the channel (Section 3.4.5). Bottom friction could affect the entire water column over shoals while altering only a portion of the water column in the channel. Because of this frictional difference from channel to shoals, tidal currents should display lateral gradients or shears (Figure 5.1c). Similarly, as tidal currents should

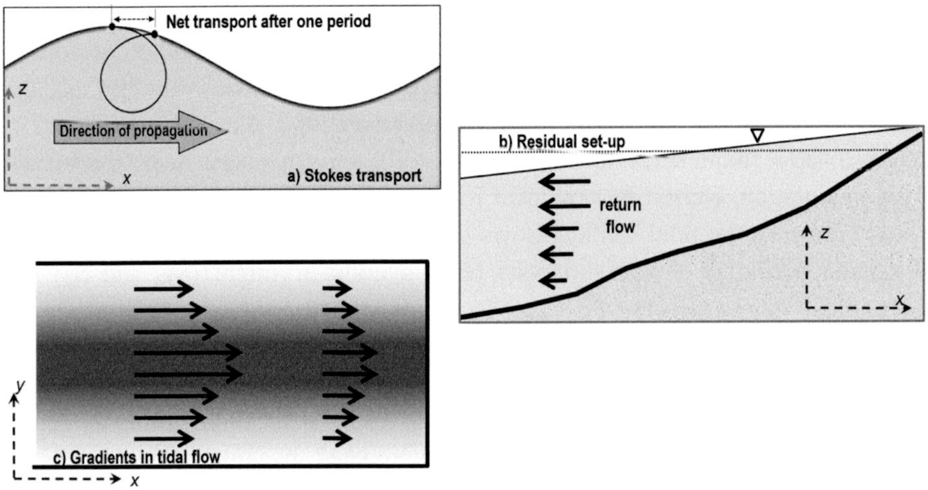

Figure 5.1 Schematic of mechanisms producing tidal residual flows in semienclosed basins. (a) Stokes transport per unit width (m^2/s) related to the covariance between water elevation and wave (tidal) velocity in an asymmetric wave, that is, arising from the average over one or more cycles ($\overline{U\eta}$). Concurrently, the distance traveled by one particle after one period is the *Stokes drift*. Both Stokes transport and drift follow the direction of wave propagation. (b) Residual set-up in water elevation resulting from the Stokes transport and the residual flow (arrows) arising from the water-level slope. (c) Advective flux of momentum (per unit mass) related to gradients in tidal flow (arrows, during flood only), which are asymmetric between flood and ebb. Dark shades represent deep areas. In this diagram, the bathymetry is invariant along the basin.

vanish at the basin's head, the currents at the entrance are expected to be stronger than at the head, resulting in along-basin gradients (Figure 5.1c). Such lateral and along-basin gradients in tidal flows will be asymmetric between flood and ebb, resulting in tidal residual flows. This mechanism is analogous to the advective flux of momentum (per unit mass) that drives overtides and compound tides. It can also be referred to as *tidal stress*.

Depending on the length of the basin, as we will see quantitatively, the residual flows will be into the basin in the channel and out of the basin over shoals, or the other way. These mechanisms are explained dynamically in Section 5.2 on the basis of a nonlinear (from the advective flux of momentum) momentum balance.

5.2 Mechanisms that Generate Tidal Residual Flows

A dynamical or quantitative explanation of the mechanisms that drive residual flows after one or more tidal cycles in a semienclosed basin is anchored in a momentum balance. The following presentation has, perhaps, numerous equations but should be straightforward to follow. With equations we save words. In this context, the depth-averaged balance explores tidal currents (along-basin U and across-basin V) in a homogeneous (constant density) semienclosed basin with no surface stress (wind) but with lateral variations in bathymetry. The dynamic balance is similar to that presented in equation 4.7, plus including flows and variations in the lateral direction y:

$$\underbrace{\frac{\partial U}{\partial t} + U\frac{\partial U}{\partial x} + V\frac{\partial U}{\partial y}}_{\textit{local and advective accelerations}} = - \underbrace{g\frac{\partial \eta}{\partial x}}_{\textit{pressure gradient}} - \underbrace{\frac{C_b U\sqrt{U^2+V^2}}{H+\eta}}_{\textit{bottom friction}} \tag{5.1}$$

$$\underbrace{\frac{\partial V}{\partial t} + U\frac{\partial V}{\partial x} + V\frac{\partial V}{\partial y}}_{\textit{local and advective accelerations}} = - \underbrace{g\frac{\partial \eta}{\partial y}}_{\textit{pressure gradient}} - \underbrace{\frac{C_b V\sqrt{U^2+V^2}}{H+\eta}}_{\textit{bottom friction}} \tag{5.2}$$

where C_b, remember, is a dimensionless bottom-drag coefficient (typically 0.0025), $\partial\eta/\partial x$ and $\partial\eta/\partial y$ are surface slopes along and across the basin, t is time, and g is the gravitational field force per unit mass. The mass balance, or continuity, that complements the momentum balance follows equation 4.1 for constant width B and including lateral flows:

$$\frac{\partial \eta}{\partial t} + \frac{\partial (H+\eta)U}{\partial x} + \frac{\partial (H+\eta)V}{\partial y} = 0. \tag{5.3}$$

As in equations 3.14 and 3.15 in Chapter 3, a Fourier decomposition of the quadratic friction term (last term in equations 5.1 and 5.2) allows the frictional term to become linear and the depth-averaged momentum balance to turn into (we have skipped some mathematical steps to avoid getting lost)

$$\frac{\partial U}{\partial t} + U\frac{\partial U}{\partial x} + V\frac{\partial U}{\partial y} = -g\frac{\partial \eta}{\partial x} - r_f\frac{U}{H} + \frac{r_f}{H^2}U\eta \tag{5.4}$$

$$\frac{\partial V}{\partial t} + U\frac{\partial V}{\partial x} + V\frac{\partial V}{\partial y} = -g\frac{\partial \eta}{\partial y} - r_f\frac{V}{H} + \frac{r_f}{H^2}V\eta, \tag{5.5}$$

where $r_f = 8C_bU_0/3\pi$ (slightly different from that appearing in equation 3.15) is a linearized friction coefficient with units m/s, and U_0 is the tidal current amplitude. Using the canonical value of C_b, the coefficient r_f can be approximated as $r_f = 0.002U_0$. The continuity equation (5.3) remains unchanged. Equations 5.3 through 5.5 can be solved analytically through a mathematical technique called *perturbation analysis*. This technique assumes that the solutions, that is, the dependent variables (U, V, and η, in this case), have dominant order of magnitude components, followed by lower order of magnitude contributions. The leading-order terms describe tidal variables U_t, V_t, and η_t. But we are concerned here with the next order of magnitude terms or tidal residuals U_r, V_r, and η_r. The main point is that the momentum balance (5.4 and 5.5) can be written for the tidal residual order as separate equations, where the overbar denotes tidal averages:

$$\overline{U_t \frac{\partial U_t}{\partial x}} + \overline{V_t \frac{\partial U_t}{\partial y}} = -g\overline{\frac{\partial \eta_r}{\partial x}} - \frac{r_f}{H}\overline{U_r} + \frac{r_f}{H^2}\overline{U_t \eta_t} \tag{5.6}$$

$$\overline{U_t \frac{\partial V_t}{\partial x}} + \overline{V_t \frac{\partial V_t}{\partial y}} = -g\overline{\frac{\partial \eta_r}{\partial y}} - \frac{r_f}{H}\overline{V_r} + \frac{r_f}{H^2}\overline{V_t \eta_t}. \tag{5.7}$$

Equations 5.6 and 5.7 can be solved directly for the residual flow $\overline{U_r}$ and $\overline{V_r}$:

$$\overline{U_r} = \underbrace{\frac{\overline{U_t \eta_t}}{H}}_{Stokes} \underbrace{-\frac{H}{r_f}\left[\overline{U_t \frac{\partial U_t}{\partial x}} + \overline{V_t \frac{\partial U_t}{\partial y}}\right]}_{tidal\ stress} - \underbrace{\frac{gH}{r_f}\overline{\frac{\partial \eta_r}{\partial x}}}_{residual\ slope} \tag{5.8}$$

$$\overline{V_r} = \underbrace{\frac{\overline{V_t \eta_t}}{H}}_{Stokes} \underbrace{-\frac{H}{r_f}\left[\overline{U_t \frac{\partial V_t}{\partial x}} + \overline{V_t \frac{\partial V_t}{\partial y}}\right]}_{tidal\ stress} - \underbrace{\frac{gH}{r_f}\overline{\frac{\partial \eta_r}{\partial y}}}_{residual\ slope}. \tag{5.9}$$

These are the equations that hold the substance to explain the mechanisms driving tidal residual flows (Figure 5.1). If it is challenging to follow the procedure from 5.1 to 5.9, do not worry about the details. The key result is contained in equations 5.8 and 5.9. From these equations, we can directly see how the residual flows $\overline{U_r}$ and $\overline{V_r}$ are produced by (all sketched in Figure 5.1) (a) Stokes transport velocities, (b) residual slope, and (c) tidal stresses or advective fluxes of tidal momentum.

Stokes transport velocities represent the covariance of tidal current and tidal elevation, normalized by water depth. These velocities will be into the basin, greatest over the shallowest water column (inversely proportional to H). But Stokes velocities can be zero or near zero if the tidal currents and elevation are in or close to quadrature (standing wave) because the covariance $\overline{U_t \eta_t}$ vanishes. Residual currents derived by tidal stresses are related to gradients in tidal currents and will be prominent over the deepest water column, as they are proportional to H. Currents derived from residual slopes will be out of the basin and also will be most pronounced in the deep channel of a cross-section because they are proportional to H.

5.3 Flows in "Short" and in "Long" Basins

The depth-averaged flow structure resulting from the three mechanisms described in equations 5.8 and 5.9 will depend on the length of the basin L as compared to

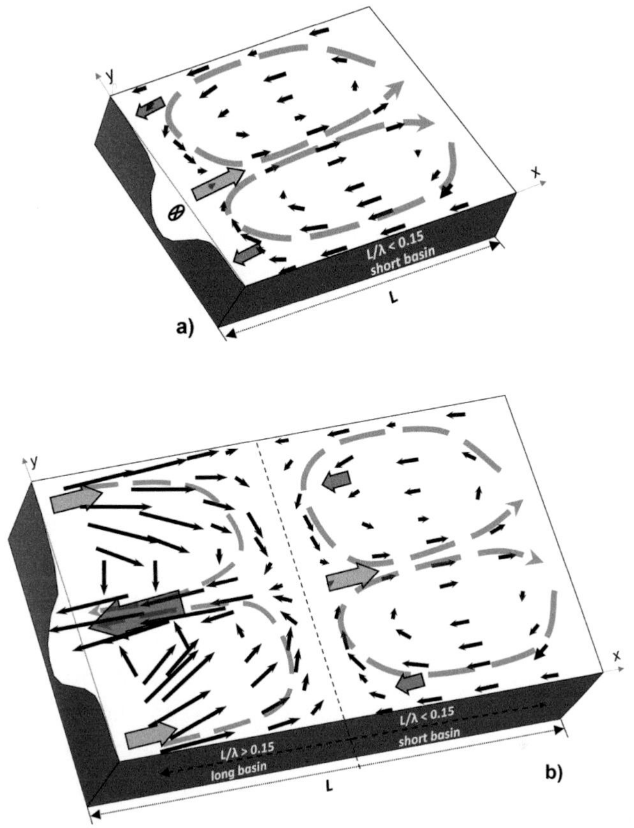

Figure 5.2 Schematic of depth-averaged tidal residual circulation for (a) short and (b) long basins with channel–shoal bathymetry (not to scale). Black arrows indicate the horizontal structure of counter-rotating circulations in the basin, highlighted by thick, dashed, maroon arrows. The net transport in channel and over shoals is illustrated by the thickest and translucent arrows. In a long basin ($L/\lambda > 0.15$), the circulation structure of a short basin appears near thc head.

the tidal wavelength λ. This is because of the distinct behavior of tidal variables (U_t, V_t, η_t) inside basins with different L. In a "short" basin, where $L/\lambda < 0.15$, that is, where the basin length is shorter than ~15% of the tidal wavelength, the tide will likely display a standing wave behavior. In this "short" basin, the depth-averaged tidal residual circulation at the entrance will display inflow in the channel and outflow over shoals (Figure 5.2a). Keep in mind that a typical tidal wavelength $\lambda(= CT = \sqrt{gH}T)$ for a semidiurnal tide over a depth of 10 m is about 400 km. So, for a depth H of 10 m, a "short" basin will be shorter than about 60 km, but this will change with the value of H. In this "short" basin, the tidal residual circulation should theoretically exhibit a pair of counter-rotating gyres inside the basin (Figure 5.2a). A key result here is that net inflow is expected in the channel, combined with outflow over shoals.

In a basin that is longer than 15% of the tidal wavelength ("long" basin) there are two pairs of counter-rotating gyres in the depth-averaged tidal residual circulation (Figure 5.2b). Near the basin's entrance, one pair of gyres describes outflow in the channel and inflow over shoals. Toward the basin's head, the second pair of gyres is the same as in a short basin. In fact, the basin responds as if it were "short" in the portion where $L/\lambda < 0.15$.

The main point of these theoretical results is that in "short" and frictional basins, depth-averaged exchange flows at the basin's entrance should exhibit tidal residual inflows in the channel and outflows over shoals. In "long" and frictional (dynamically shallow) basins, exchange flows at the entrance will display the opposite structure with outflow in the channel and inflow over shoals. Limitations of those theoretical results include neglect of the effect of the Earth's rotation (*Coriolis* accelerations), disregard of the vertical structure of the residual currents, uniform bathymetry along the basin, as well as straight and parallel along-basin coastlines.

With respect to *Coriolis* accelerations, they could become relevant in dynamically deep basins where frictional effects are negligible (see Section 3.4.3). The relevance of friction can be scaled by comparing a frictional term $r_f\overline{U_r}/H$ (e.g., equation 5.6) to a scaled version of *Coriolis* accelerations $f\overline{U_r}$, where f is the *Coriolis* parameter (equation 2.34). The comparison yields a nondimensional number:

$$E = r_f/fH, \tag{5.10}$$

which can be approximated as

$$E = 0.002\ U_0/fH, \tag{5.11}$$

where the 0.002 coefficient is $8C_b/3\pi$. In equation 5.10, the ratio r_f/f represents a depth of frictional influence (in meters). Therefore, E compares the depth of frictional influence to total water column depth, analogous to the Ekman number (described in Chapter 7). When $E < 0.1$, *Coriolis* acceleration should dominate over friction (darkest shade in Figure 5.3, for values of $\log(E) < -1$) and the above theoretical results become invalid. This is expected to occur under weak tidal currents U_0 (< 0.1 m/s) and depths > 10 m, for example in fjords. In the range $-1 < \log(E) < 0$ *Coriolis* accelerations probably play a role in the dynamics, together with friction (light shade in Figure 5.3). The white shaded region in Figure 5.3, where $\log E > 0$, illustrates the combinations of tidal current amplitudes and depths for which the above theoretical results apply.

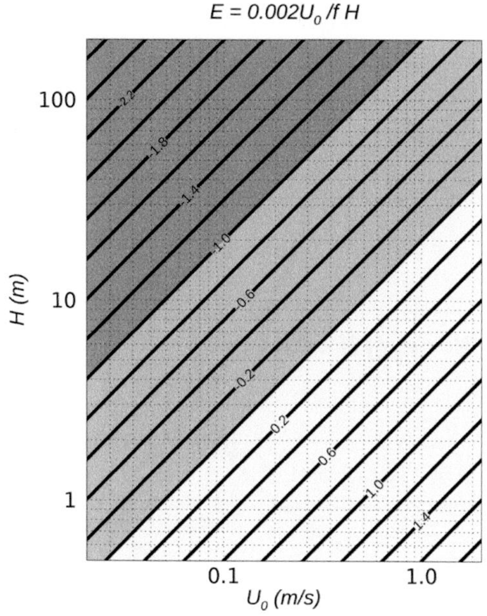

Figure 5.3 Values of the nondimensional number log (E) (equation 5.11) for different values of tidal current amplitudes U_0 and depths H. Values of E are plotted as its logarithm because negative values indicate that the denominator in equation 5.11 is greater than the numerator. Both axes are in logarithmic scale.

For basins in the shaded regions (both shades) of Figure 5.3, where tidal currents are relatively weak and depth is relatively large, the vertical and lateral structure of tidal residual currents becomes relevant. In those instances, and through a mathematical development that is beyond the scope of this text, tidal residual flows may become two-layered or even three-layered, as in fjords. In the darkest shades of Figure 5.3, tidal flow is essentially frictionless, as in fjords. In these instances where the basin's width is much smaller than the Rossby radius (Section 3.4.4), tidal residual flows are expected to be two-layered or even three-layered (Figure 5.4). Examples of this tidal residual flow structure in fjords and of the theoretical results related to equations 5.8 and 5.9 are provided in the following subsection.

5.4 Observational Examples

The first example that illustrates the qualitative agreement between observations and theoretical expectations from equations 5.8 and 5.9 (as per Figure 5.2) is drawn from the St. Augustine inlet in northeast Florida. The system was exclusively forced by tides during the day of measurements throughout one

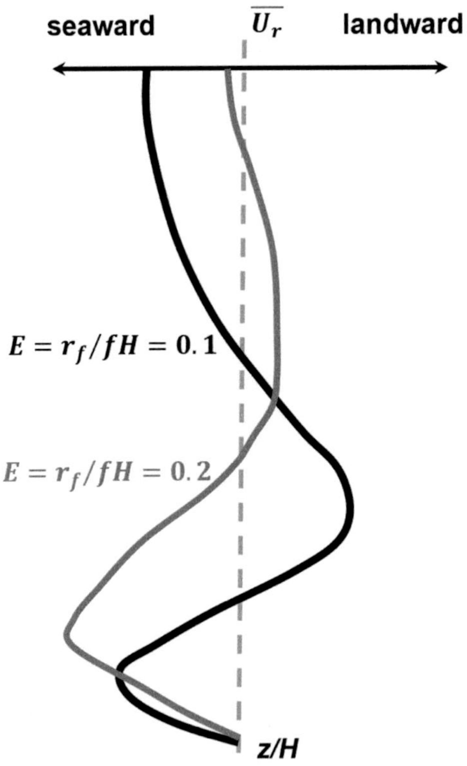

Figure 5.4 Schematic of expected theoretical profiles of tidal residual flow in relatively deep basins where E is 0.1 and 0.2.

semidiurnal tidal cycle along a cross-inlet transect. The semienclosed basin sampled was practically homogeneous. Velocity profiles were collected underway with an acoustic Doppler current profiler (ADCP) on February 2, 2006. Each point of the transect provided time series with 13 values of velocity and time, corresponding to each one of the transect occupations. Each one of the time series was processed, as in Appendix 4.1, to extract the residual flow, which is drawn in Figure 5.5. The tidal residual flow shows inflow in the channel and outflow over shoals, as expected from theory for a "short" basin. In fact, the basin displays standing-wave behavior as the water elevation and currents are 90° out of phase.

Similar behavior to the St. Augustine Inlet (Figure 5.5) is drawn from Ensenada de La Paz, in the southern Baja California peninsula on the side of the Gulf of California. In that case, the tide is mixed with semidiurnal dominance and behaves as a standing wave, that is, a "short" basin. The data-collection experiment extended for 24 hours. Also, the bathymetry displays two channels, separated by a shoal. One of the channels, however, is shallower than the other. In the experiment with a towed ADCP, the water was also homogeneous, so the main driver of

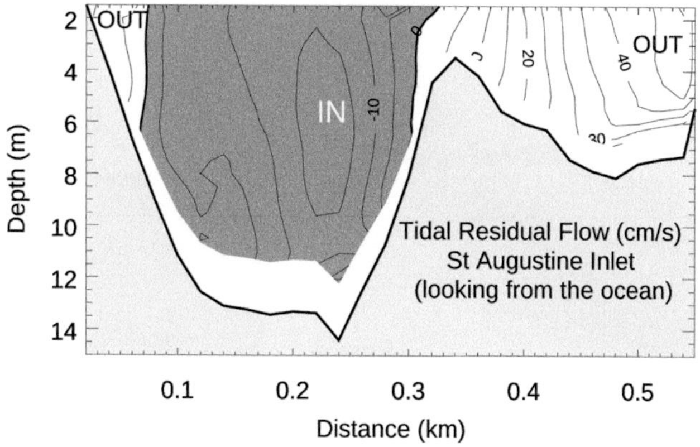

Figure 5.5 Tidal residual flows at St. Augustine Inlet, Florida, after one tidal cycle. Negative values (shaded) describe inflows. View is into the basin. Bottom profile is portrayed by the thick black line. The white masked region above the bottom is the region affected by side-lobe effects, where data are unusable.

residual flows was the tides. The example shows results for spring and neap tides (Figure 5.6a and b) and illustrates tidal residual inflow in the deepest channel and outflow in the other shallower channel. The residual flows are stronger during spring tides, compared to neap tides, indicating that the main driver of these residuals is indeed the tides.

Tidal modulation of residual flows is confirmed with a 60-day record of velocity profiles (Figure 5.6c) in the channel that displays outflow in the surveys (western channel of the middle transect in Figure 5.6a, b). The residual flows in Figure 5.6c were obtained from filtering out tidal variations. Residual flows were positive (outflow) throughout the period of observations, consistent with the surveys (Figure 5.6a, b). Weak inflow at the surface developed near the surface during the neap tides of days 99 and 114. But the most remarkable feature of this record is that the strongest outflows coincided with spring tides. These results were also consistent with theoretical arguments.

The final example of residual flows in a "short" basin displays the evolution of tidal residual flow profiles over one month in San Quintin Bay. This is a subtropical semiarid lagoon on the Pacific side of the northern coast of the Baja California peninsula, Mexico. Tides feature a standing wave behavior, typical of "short" basins. Velocity profiles were recorded in a channel, flanked by shoals, at the entrance to the lagoon. Based on theory, the residual flow expected at the site was inflow. Indeed, the observed tidal residual flow was into the lagoon (Figure 5.7) throughout the month of observations. The strongest inflows coincided with spring tides and the weakest inflows appeared in neap tides, in agreement with the record in southern Baja California (Figure 5.6c). Observational examples at the entrance to other semienclosed

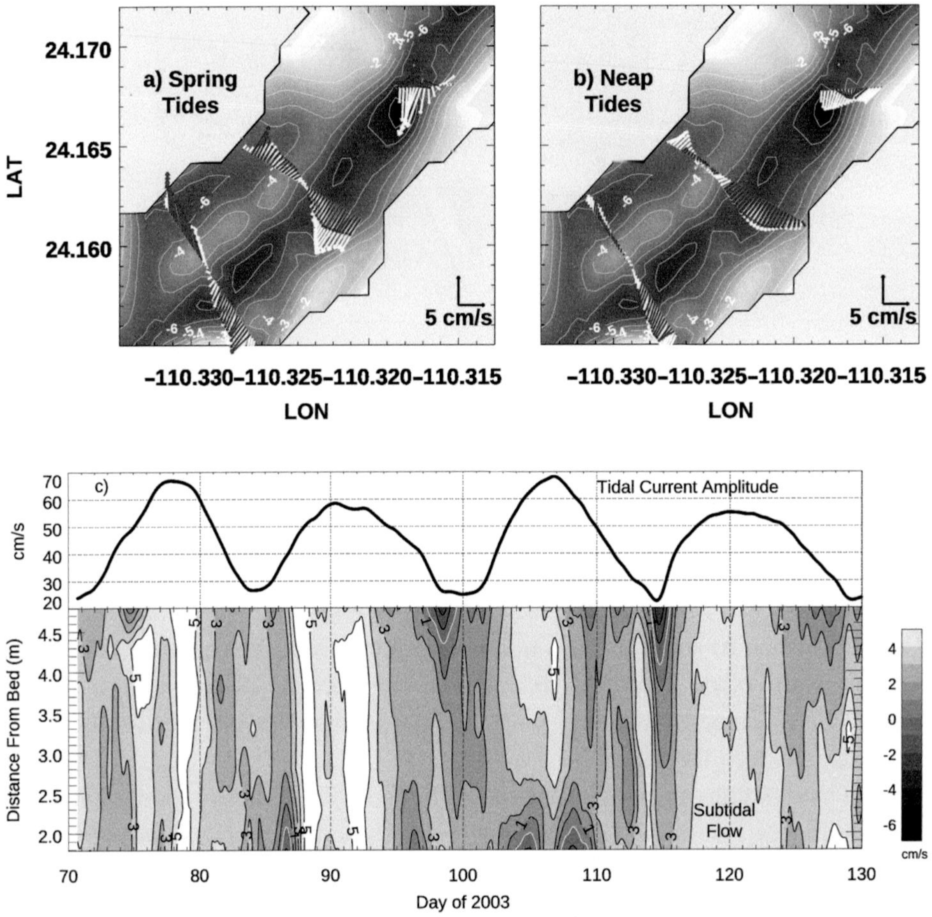

Figure 5.6 Tidal residual flows at the entrance to Ensenada de La Paz, Baja California Sur. The top two panels show tidal residual flows at three transects sampled during 24-h surveys. White arrows represent surface currents while gray arrows are near-bottom currents. Inflow develops in the relatively deeper eastern channel and outflow in the western channel. Bathymetry is displayed in shaded contours, labelled in meters. The lower panel illustrates along-channel velocity profiles, after tides are removed, over two months of observations of the western channel. Periods of spring and neap tides are portrayed by the tidal current amplitude line. Velocity contours are overwhelmingly positive (outflow) and modulated in such a way that the strongest outflows appear in the four spring-tide periods. Weakest outflows (darker contours) occur in the four neap tides.

basins, forced mainly by tides, would reinforce the concept that tidal residual flows will display inflow in channels and outflow over shoals.

There are few observational examples of tidal residual flows in "long" basins. This one comes from the entrance to the Chilean Inland Sea, where tidal currents

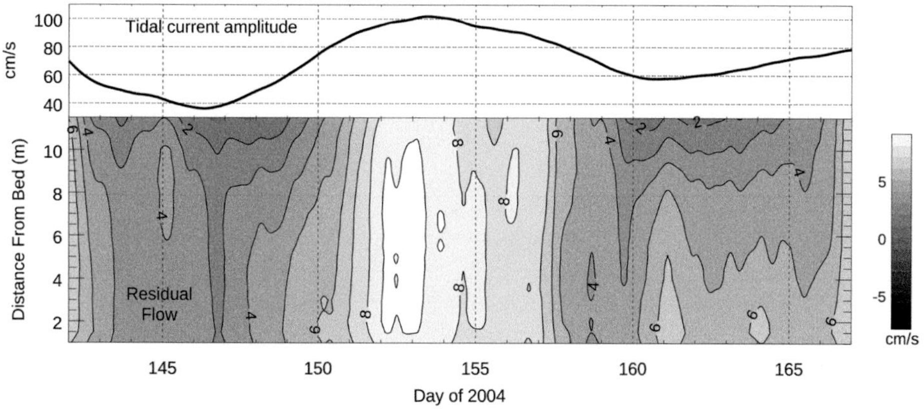

Figure 5.7 Tidal residual flows in San Quintin Bay, Baja, California. The plot has the same features as Figure 5.5c. Tidal current amplitude shows two neap periods (days 146 and 160) and one fully reached spring tides (days 152–153). Positive values in the contour plot indicate inflow, which occurs throughout the period of observations. Greatest inflow appeared on day 152–153, coinciding with spring tides.

can reach up to 3.5 m/s through one of the channels that connects the inland sea with the Pacific Ocean. Chacao channel reaches depths of 100 m but the strength of the tidal currents makes it a dynamically shallow basin, following Figure 5.3 and equation 5.11 where E ~ 0.7. Tides behave as progressive waves and the residual flow across this channel exhibits inflow preferentially over shoals and outflow in channel (Figure 5.8), as predicted by theory for "long" basins. In the Chilean Inland Sea, the ratio L/λ lies between 0.2 and 0.4.

When tidal flows are unaffected by lateral changes in bathymetry the basin is dynamically deep as friction is negligible. The tidal residual flow in this situation can change direction with depth. Such is the following example from a fjord, also in the Chilean Inland Sea. Reloncavi fjord is >100 m deep. Tidal cycle measurements with a towed ADCP provided a cross-section of tidal residual flows (Figure 5.9) that displayed marked depth-dependence. There was a surface layer outflow that in this case was enhanced by buoyant discharge in the fjord. Underneath, there was an intermediate layer inflow that can be interpreted as density driven. However, theoretical considerations can explain this inflow layer. Further underneath, there was another outflow layer. For these types of flows to develop, the water column has to be sufficiently deep to provide a frictionless tidal flow (consistent with Figure 5.4). The theory is well developed to explain these depth-dependent tidal flows. More examples are necessary to fully validate the theory.

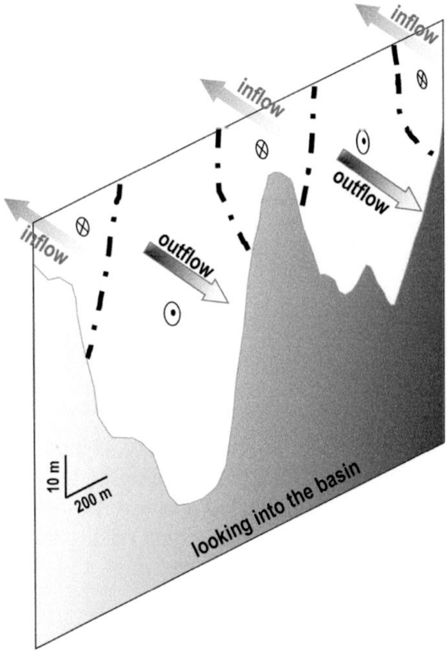

Figure 5.8 Schematic of tidal residual flows across Chacao Channel at the entrance to the Chilean Inland Sea. Tidal residual inflows appear over shoals and outflow in the channel, as expected from theory in a "long" basin.

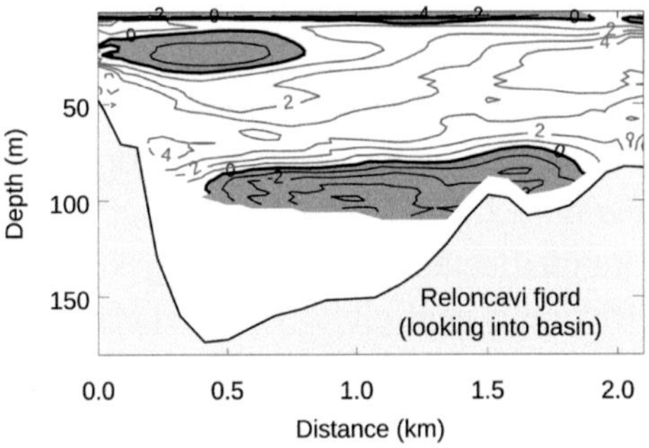

Figure 5.9 Tidal residual flows in Reloncavi fjord, Chile. The view is into the fjord, with shaded contours indicating tidal residual outflow. The flow structure illustrates a three-layered residual circulation. Measurements with the ADCP extended to depths of ~100 m without reaching all the way to the bottom in portions of the transect.

5.5 Flows around Headlands

Tidal residual flows can also arise from tidal flows that are modified by prominent coastline irregularities such as points or headlands (e.g., Figure 5.10). Modifications on each side of the headland are asymmetric from one phase of the tide to the

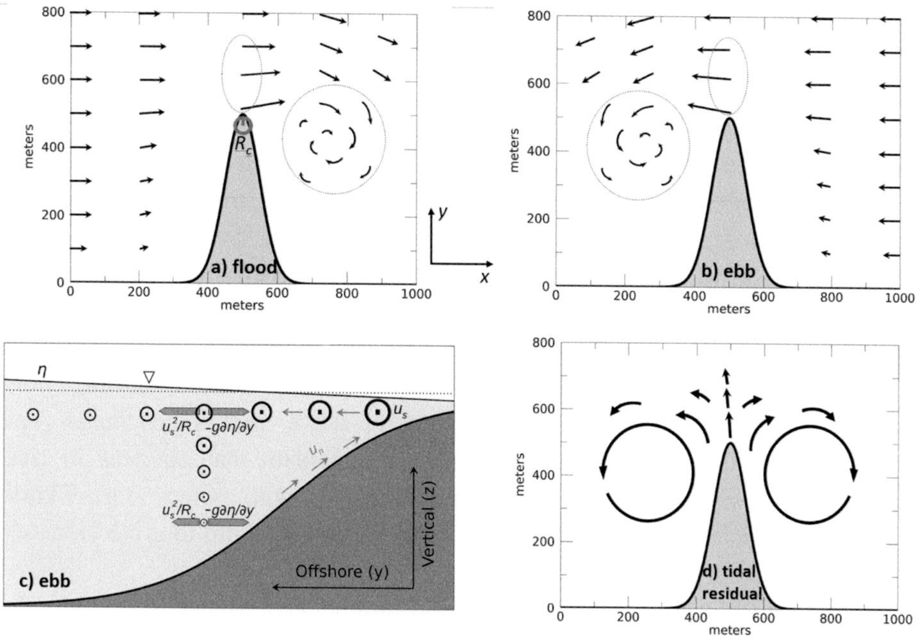

Figure 5.10 Conceptual diagram of tidal and residual flows around a headland of radius of curvature R_c. A clockwise eddy develops on the lee of the headland (a) when the headland is to the right of the tidal flow direction. Labelling this panel as "flood" is arbitrary and for reference purposes only. A counterclockwise eddy appears on the lee of the headland (b) when the tidal flow reverses direction, that is, when the headland is to the left of the flow direction. A cross-section (depth vs distance) off the headland shows the acceleration of the streamwise flow u_s, denoted by circles with dots in the middle, and the development of secondary or cross-shore flows away from the headland at the surface and toward the headland underneath. The stream-normal momentum balance, consisting of centrifugal accelerations balanced by pressure gradient force, is shown in thick arrows, with an imbalance in favor of the pressure gradient force deeper in the water column. This imbalance drives the stream-normal near-bottom flow toward the headland (upwelling).

other, resulting in non-tidal flows on each side. Within a tidal cycle, tidal flow is expected to accelerate as it goes around a headland of radius R_c (Figure 5.10a). Downstream of the headland the lateral gradients (lateral shears) of the flow associated with near-stagnant conditions on the protected side of the headland favor a transient eddy. The eddy is clockwise when the headland is on the right of the flow direction (e.g., region within the dotted circle in Figure 5.10a). It follows that the eddy describes counterclockwise circulation when the headland is to the left of the flow direction (inside the dotted circle in Figure 5.10b).

Focusing on the processes in front of the headland, the acceleration of the streamwise flow u_s causes a region of relatively low pressure (inside the dotted oval of Figures 5.10a, b), according to basic physical principles (e.g., Bernoulli's principle). Relatively lower pressure off the headland manifests itself in a drop of water level η and an offshore pressure gradient ($g\partial\eta/\partial y$, Figure 5.10c). The acceleration of u_s caused by curvature at the headland drives a stream-normal flow away from the headland at the surface (u_n, gray vectors in Figure 5.10c), consistent with the offshore increase of η. At the surface, the stream-normal momentum balance is between centrifugal acceleration (u_s^2/R_c) and the offshore pressure gradient (thick gray arrows in Figure 5.10c). Bottom friction weakens the streamwise flow (dot-circles in Figure 5.10c) causing an imbalance near the bottom in favor of the pressure gradient force (per unit mass), which drives a near-bottom flow toward the headland (Figure 5.10c). The resulting stream-normal flow u_n, consisting of near-surface offshore flow and near-bottom onshore flow, provides a mechanism of upwelling related to flows around headlands. This "secondary circulation" represents the same mechanism that operates in river meanders and a possible mechanism that competes with others (e.g., Ekman dynamics associated with wind stress and *Coriolis* acceleration) to drive increased primary productivity around headlands.

An example of the secondary circulation around a headland is provided by ADCP measurements obtained off Cape Henry at the entrance to Chesapeake Bay (Figure 5.11). At two different times of an ebb phase, the stream-normal flow displays motion away from the cape at the surface and toward the cape underneath. It is possible that this circulation provides nutrients to the surface waters in this portion of Chesapeake Bay, known as feeding grounds for birds and mammals and as a retention region of larvae of commercial importance.

The tidally averaged flow arising from the interaction of oscillatory flows with a headland consists of a pair of counter-rotating eddies (Figure 5.10d). This clearly results from the average of the flow distributions during flood (Figure 5.10a) and ebb (Figure 5.10b). Looking offshore from the base of the headland, a clockwise recirculation is expected to develop to the right of the headland, and a counterclockwise recirculation to the left. This tidal residual circulation has key implications for water quality and sediment transport. The central region of these recirculations may become relatively stagnant and allow deposition of suspended materials and accumulation of suspended solutes.

An example of tidal residual circulation around a headland was derived from one month of high-frequency radar surface velocity measurements in the Gulf of Fonseca, Central America. The gulf is shared by El Salvador, Honduras, and Nicaragua, on the Pacific coast. The month-long average of surface currents illustrates near-surface offshore flow from Point Cosigüina, the prominent headland in Nicaragua (Figure 5.12). Residual surface flows are directed offshore, flanked by a partially developed pair of counter-rotating eddies. Eddies are not completely apparent because of obstructions by coastlines and islands. The flow

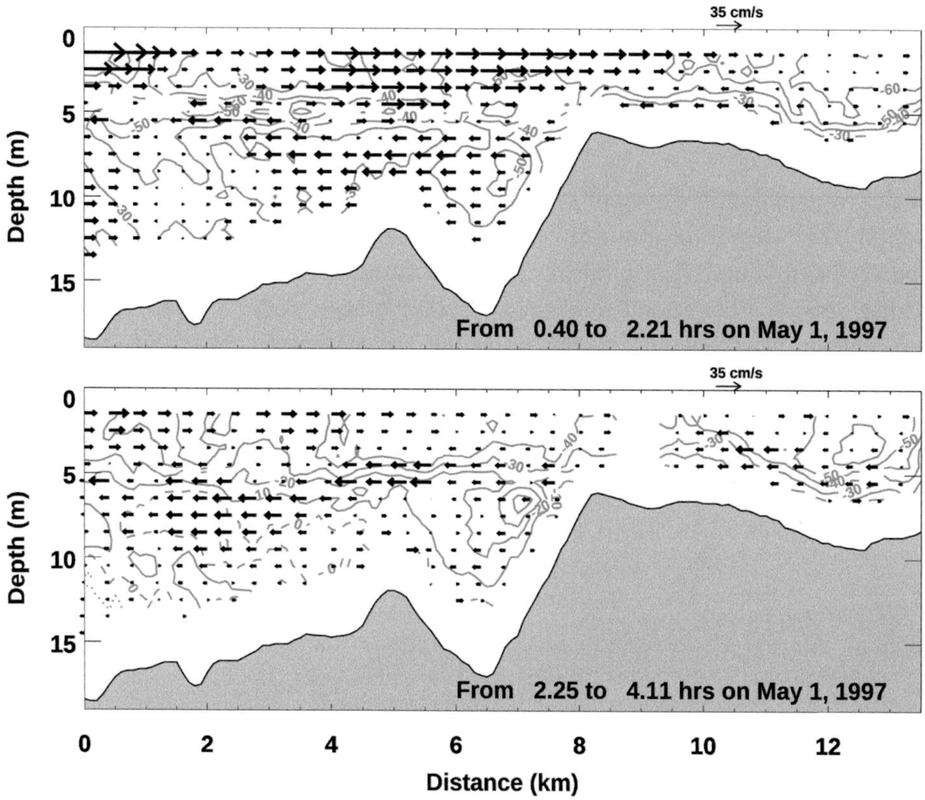

Figure 5.11 Example of secondary circulation at the entrance to Chesapeake Bay during two different repetitions of the same transect during ebb tides. View is into the basin with Cape Henry on the left. Gray continuous contours indicate the flow perpendicular to the viewer (in cm/s, coming out of the paper). Arrows illustrate the secondary circulation, mainly in the first 7 km of the transect, with flow away from the cape near the surface and toward the cape underneath. The gray shade is the bottom profile.

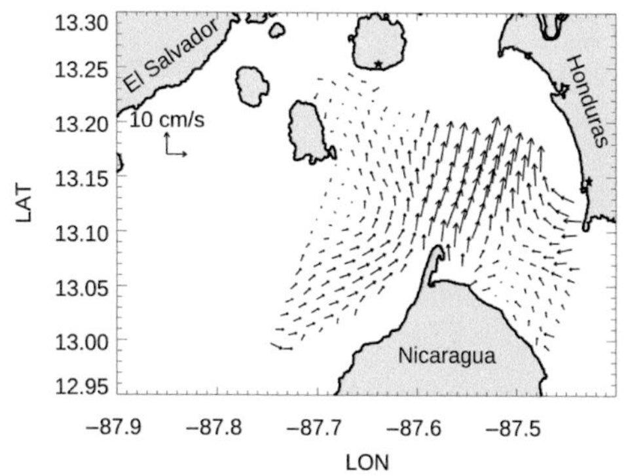

Figure 5.12 Example of tidal residual circulation off a headland on the Nicaraguan coast (Point Cosigüina) in the Gulf of Fonseca, Central American Pacific. Flow field was generated from a month-long deployment of two high-frequency radar antennae marked by stars on the figure.

structure suggests that upwelling should develop off the headland because of the surface flow divergence there.

5.6 Take-Home Message

Tidal residual flows in semienclosed basins result mainly from Stokes transport, return flow from the residual slope caused by Stokes transport, and from asymmetries in tidal flow gradients. The tidal residual circulation at the entrance to "long" basins is expected to display depth-averaged outflow in a channel and inflow over the shoals that flank the channel. At the entrance to a short basin, the tidal residual is expected to display inflow in channel and outflow over shoals. Around headlands, tidal flows may trigger upwelling of subsurface waters. Tidal residual flows are expected to describe a pair of counter-rotating gyres at both sides of a headland.

Additional Sources

Berthot, A., and C. Pattiaratchi (2006) Mechanisms for the formation of headland-associated linear sandbanks. *Cont. Shelf Res.* 26(8): 987–1004.

Geyer, W.R. (1993) Three-dimensional tidal flow around headlands. *J. Geophys. Res.* 98 (C1): 955–966.

Ianniello, J.P. (1977) Tidally induced residual currents in estuaries of constant breadth and depth. *J. Mar. Res.* 35: 755–786.

Ianniello, J.P. (1981) Comments on tidally induced residual currents in estuaries: Dynamics and near bottom flow characteristics. *J. Phys. Ocean.* 11: 126–134.

Li, C.Y., and J. O'Donnell (2005) Tidally driven residual circulation in shallow estuaries with lateral depth variation. *J. Geophys. Res.* 102(C13): 915–927, 929.

Signell, R.P., and W.R. Geyer (1991) Transient eddy formation around headlands. *J. Geophys. Res.* 96(C2): 2561–2575.

Winant, C.D. (2008) Three dimensional tidal residual circulation in an elongated, rotating basin. *J. Phys. Ocean.* 38: 1278–1295.

6
Wind-Driven Flows in Homogeneous, Semienclosed Basins

This chapter continues studying flows in homogeneous semienclosed basins. The main flow driver in this case is the wind. Density gradients and their dynamic impact continue to be neglected. Flows driven by density gradients will be treated in the next chapter (Chapter 7). The present chapter begins with a presentation of fundamental concepts related to wind-driven flows in homogenous fluids. Then it goes into a quantitative presentation of the fundamental dynamics related to these flows in semienclosed basins. It describes the effects of lateral variations in bathymetry on wind-driven flows. The chapter concludes with a comparison of theoretical results to observations of wind-driven flows.

6.1 Fundamental Concepts

As wind blows on the surface of a basin, it transfers momentum from the air to the water (see also Section 2.6). Horizontal wind velocity decreases vertically because of the air–water boundary. This vertical gradient, or vertical shear, in wind velocity provides a vertical transfer of horizontal momentum from the air to the water (Figure 2.6). In addition to deforming the interface and causing waves, which are studied comprehensively in other texts, winds can produce three main responses in semienclosed basins: movement of surface waters in the direction of the wind, setting-up of water-level slopes in the same wind direction, and driving a vertical flux of momentum through turbulent stresses (mixing of momentum). This chapter concentrates on wind-driven currents and water-level slopes, as details associated with the mixing of momentum require a more advanced treatment that involves turbulence through Reynolds stresses (see Sections 2.5 and 2.6).

As the wind drags surface waters downwind, waters pile up against a downwind boundary and establish a water-level slope. This slope, in turn, drives upwind flow (against the wind). Such upwind flow may develop either (i) underneath the surface flow or (ii) throughout all or most of the water column in relatively deeper

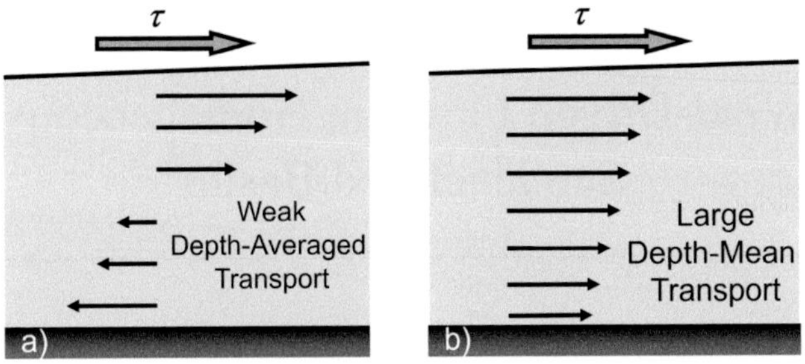

Figure 6.1 Diagram of possible response of a water column to wind stress. The response may be through a two-layer exchange flow with downwind flow at surface and upwind flow underneath, or downwind flow throughout the water column. The conditions that favor either response are still undeciphered. The rest of this chapter proposes an attempt to begin untangling such a response.

bathymetric conduits. A still unresolved question is: Under what circumstances does wind drive unidirectional flows throughout the water column and when is the flow bidirectional (e.g., Figure 6.1)? This chapter offers possible answers to such question following a quantitative perspective. The answers in this chapter, however, are restricted to homogeneous fluids.

6.2 Basic Physics

To study along-basin (u) wind-driven flows in semienclosed, homogeneous systems, we take the along-basin (x) momentum balance (from the first equation 2.40):

$$\underbrace{\frac{\partial u}{\partial t}}_{local} \underbrace{+u\frac{\partial u}{\partial x}+v\frac{\partial u}{\partial y}+w\frac{\partial u}{\partial z}}_{advective} - \underbrace{fv}_{Coriolis} = \underbrace{-g\frac{\partial \eta}{\partial x}}_{pressure\ gradient} \underbrace{+\frac{\partial}{\partial z}\left[A_z\frac{\partial u}{\partial z}\right]}_{friction}, \tag{6.1}$$

where w is the velocity component in the vertical z direction (negative down the water column), v is the component in the lateral direction y, g is gravity's acceleration, f is the *Coriolis* parameter, η is water elevation, and A_z is the vertical eddy viscosity. Equation 6.1 implies that vertical frictional effects dominate horizontal friction, that is, that $A_z L_x^2/A_h H^2 \gg 1$, where L_x is the basin's length and A_h is the horizontal eddy viscosity (see Section 2.6). To explore wind-driven flows, we further simplify equation 6.1 under the assumptions of linear, non-rotational, and steady motion. This makes the left-hand side of equation 6.1 equal to zero. We also assume that A_z and the water-level slopes ($\partial\eta/\partial x$) in the basin are constant, which results in the following momentum balance:

$$0 = \underbrace{-g\frac{\partial \eta}{\partial x}}_{\textit{pressure gradient}} + \underbrace{A_z\frac{\partial^2 u}{\partial z^2}}_{\textit{friction}}. \tag{6.2}$$

Equation 6.2 is an ordinary differential equation where the dependent variable u may be solved as a function of the independent variable z. The solution can be obtained by integrating equation 6.2 twice with respect to z. If the procedure to arrive at the solution is unimportant, skip to equation 6.8. However, the procedure is straightforward. Integrating equation 6.2 once, results in

$$\frac{\partial u}{\partial z} = \frac{g}{A_z}\frac{\partial \eta}{\partial x} z + c_1, \tag{6.3}$$

where $c_1$is a constant of integration. Integrating again, yields

$$u(z) = \frac{g}{2A_z}\frac{\partial \eta}{\partial x} z^2 + c_1 z + c_2. \tag{6.4}$$

This is the general solution to the momentum balance in equation 6.2. The strength of the wind-driven flow depends directly on the slope $\partial\eta/\partial x$ and inversely on the eddy viscosity A_z. Equation 6.4 indicates that the wind-induced flow will have a parabolic profile with depth z because of its dependence on z^2. There are two constants of integration, c_1 and c_2, whose values are obtained from implementing two boundary conditions. For the first boundary condition, we assume that the vertical gradient (or vertical shear) at the surface ($z = 0$) is provided by the wind stress (also equation 2.41), that is,

$$A_z\frac{\partial u}{\partial z}[z = 0] = \frac{\tau_s}{\rho_0}, \tag{6.5}$$

ρ_0 being a reference water density (e.g., 1,025 kg/m^3; the actual value of ρ_0 has negligible weight on the final result as it makes little difference to divide a number by, say, 1,010 or by, say, 1,028). With condition 6.5, the first constant of integration becomes, from equation 6.3,

$$c_1 = \frac{\tau_s}{\rho_0 A_z}. \tag{6.6}$$

For the second boundary condition we take that the flow u at the bottom ($z = -H$) is zero. This is called a *no-slip* condition. Applying the boundary condition to equation 6.4, with equation 6.6, yields the second constant of integration:

$$0 = \frac{gH^2}{2A_z}\frac{\partial \eta}{\partial x} - \frac{\tau_s H}{\rho_0 A_z} + c_2. \tag{6.7}$$

The solution $u(z)$ derived from equation 6.4 with the two constants of integration is

$$u(z) = \frac{g}{2\mathbf{A}_z}\frac{\boldsymbol{\partial\eta}}{\boldsymbol{\partial x}}[z^2 - H^2] + \frac{\boldsymbol{\tau}_s}{\rho_0\mathbf{A}_z}[z + H]. \tag{6.8}$$

We now see that the wind-driven flow represented by equation 6.8 has two contributions: one from the water-level slope $\partial\eta/\partial x$ and one from the wind stress τ_s. The slope causes flow profiles with parabolic shape because its role depends on z^2. In fundamental fluid mechanics this has the shape of a *Poiseuille flow*. In turn, the wind stress itself drives flow with a linear profile because of its dependence on z, reminiscent of a *Couette flow*. Plotting equation 6.8 needs a prescription of three parameters (free parameters in bold): the slope, the eddy viscosity, and the wind stress. For a given A_z, the parameters τ_s and $\partial\eta/\partial x$ in the solution need to be dynamically consistent, that is, the slope depends on the stress. Therefore, we need to relate the water-level slope to the wind stress. One way of finding this relationship is to assume that the transport, per unit of basin width, produced by the wind-driven flow u is zero:

$$\int_{-H}^{0} u(z)dz = 0. \tag{6.9}$$

This is the first condition, out of four presented in this chapter, that relates wind velocity profile to transport. Condition 6.9 implies that the same amount of water moved downwind by the wind stress equals the amount moved upwind. Applying equation 6.9 to 6.8, that is, integrating equation 6.8 with respect to depth, evaluating the result between $-H$ and 0 and equating to zero, results in:

$$\frac{\boldsymbol{\partial\eta}}{\boldsymbol{\partial x}} = \frac{3}{2}\frac{\tau_s}{\rho_0 g H}. \tag{6.10}$$

Equation 6.10 provides a dynamically consistent relationship between surface slope and wind stress. It says that the slope is proportional to the wind stress and inversely proportional to the water depth H. Equation 6.10 implies no *net* transport of water by the wind and can be used to obtain a solution of wind-driven flow in terms of the wind stress and the eddy viscosity. But before exploring such a solution, we explore the reliability of equation 6.10 with data (Figure 6.2a).

Data from the Chilean Inland Sea at Meninea Strait show that measured water-level slopes followed closely the scaled wind stress, as in equation 6.10, when the wind stress was southward (negative). Positive slopes were greater than those predicted by equation 6.10 most likely because the anemometer was semi-sheltered from northward (positive) wind stresses. This sheltering caused northward wind

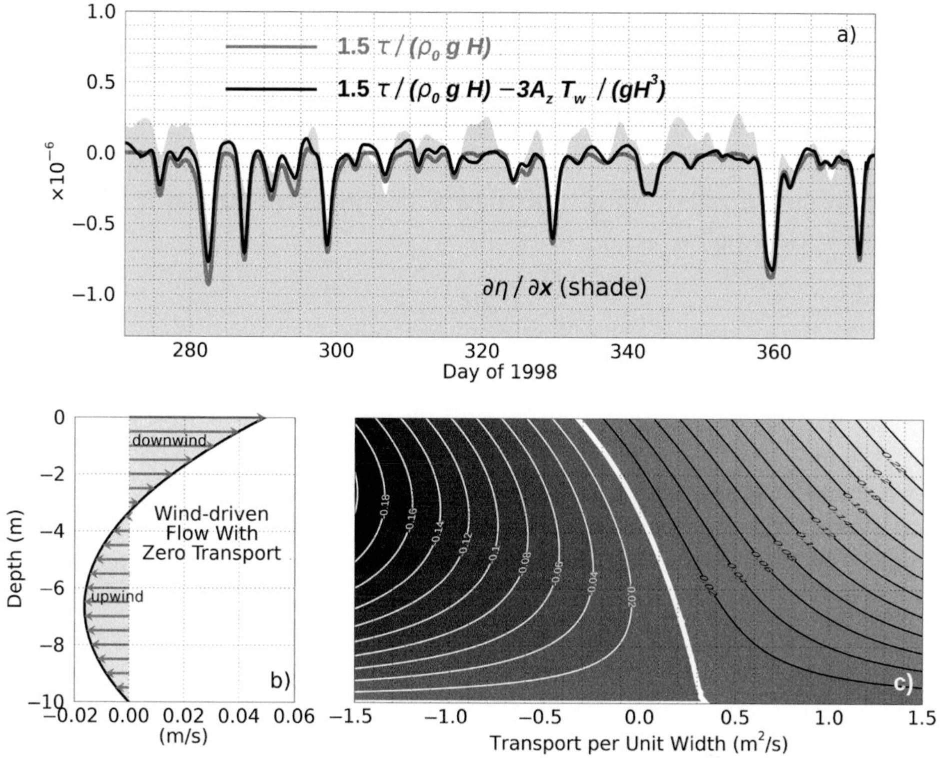

Figure 6.2 (a) Water-level slope calculated with equations 6.10 (gray line) and 6.13 (black line), compared to observed slope (shaded). Data are from 100-day records of water level, wind velocity, and water velocity profiles at Meninea Strait in the Chilean Inland Sea. (b) Wind-driven velocity profile derived from equation 6.11 for $H = 10$ m, $\tau_s = 0.1$ Pa and A_z of 0.005 m^2/s. (c) Contours of wind-driven velocity profiles for transports T_x (per unit width) ranging from -1.5 (against the wind) to 1.5 (with the wind) m^2/s. The wind stress, the vertical eddy viscosity, and the vertical axis are the same as in (b). Positive velocity values are downwind.

stresses to be underestimated in the observations. Thus, equation 6.10 can be used in a wide range of practical applications.

Returning to equation 6.8, if we substitute the water-level slope from equation 6.10, we obtain

$$u(z) = \frac{\tau_s}{\rho_0 A_z}\left[\frac{3}{4}\left(\frac{z^2 - H^2}{H}\right) + z + H\right]. \tag{6.11}$$

The velocity profile related to this equation is displayed in Figure 6.2b for specified values of τ_s, A_z, and H. This profile provides a parabolic distribution of wind-driven flow with depth. The flow strength is proportional to the wind stress τ_s and water depth H, and inversely proportional to the eddy viscosity A_z. The theoretical profile portrays downwind flow at the surface, over roughly 0.3 of the water column, and weaker upwind flow underneath. This results from the condition of no net transport (equation 6.9). A Matlab script (provided) allows specification of any desired combination of parameters to draw a velocity profile.

Next, instead of assuming a no-net transport condition (equation 6.9), a second way of relating τ_s and $\partial\eta/\partial x$ is to consider a net wind-driven transport T_x per unit of basin's width (m^2/s):

$$\int_{-H}^{0} u(z)dz = T_x. \tag{6.12}$$

This transport may be generated remotely and may be independent of τ_s. In some cases, however, the transport will depend on the wind stress. The relationship between local wind stress and remote forcing (mass transport from a region beyond the zone affected by wind) is an area of study that remains unresolved. If the transport T_x can be prescribed independently of the wind stress τ_s, applying condition 6.12 to equation 6.8 provides another way to relate the water-level slope $\partial\eta/\partial x$ to τ_s, and also to T_x:

$$\frac{\partial\eta}{\partial x} = \frac{3}{2}\frac{\tau_s}{\rho_0 g H} - \frac{3A_z}{gH^3}T_x. \tag{6.13}$$

This equation is the same as equation 6.10 for $T_x = 0$. It can be used to estimate water-level slopes with prescribed values of A_z and measurements of wind velocity and current velocity profiles that provide τ_s and T_x. This was done with the data from the Chilean Inland Sea displayed in Figure 6.2a, which also included time series of ADCP profiles. These profiles allowed calculation of T_x (by integrating velocity profiles vertically). Water-level slopes calculated with equation 6.13 improved the estimates from equation 6.10. However, in the example of Figure 6.2, the issue remains of under-predicting positive $\partial\eta/\partial x$ values because of sheltering of the meteorological station from southerly winds.

Inserting equation 6.13 into equation 6.8, we obtain a profile of wind-driven flow modified by a remotely generated transport T_x:

$$u(z) = \frac{\tau_s}{\rho_0 A_z}\left[\frac{3}{4}\left(\frac{z^2 - H^2}{H}\right) + z + H\right] - \frac{3}{2}\frac{T_x}{H}\left(\frac{z^2}{H^2} - 1\right). \tag{6.14}$$

Equation 6.14 is plotted in Figure 6.2c for transport values T_x between $-1.5\ m^2/s$ (against the wind) and $1.5\ m^2/s$ (downwind). A total of 61 profiles of equation 6.14 are represented as a contour plot where the profile at $T_x = 0$ is identical to that of Figure 6.2b. Velocity profiles become unidirectional with depth and downwind when $T_x > 0.4\ m^2/s$. Similarly, velocity profiles become unidirectional upwind when $T_x < -0.35\ m^2/s$. These threshold values of T_x that suppress bidirectional wind-driven exchange flow will ultimately depend on τ_s, A_z, and H. Two other ways of relating τ_s and $\partial\eta/\partial x$ consider lateral variations in bathymetry, that is, involve $H(y)$ instead of constant H.

6.3 Influence of Lateral Variations in Bathymetry

This is an opportune point to remember that we are trying to understand wind-driven slopes and circulation in semienclosed basins, derived from a simplified, linear momentum balance. The balance is between pressure gradient and friction, under homogeneous fluid and eddy viscosity. Expanding equation 6.9 to be valid throughout a cross-section of width B, instead of at a local profile (as in the previous section), the condition needed to relate τ_s and $\partial\eta/\partial x$ now reads,

$$\int_0^B \int_{-H}^0 u(y,z)dz\,dy = 0. \tag{6.15}$$

This condition says that there is no net transport (in m^3/s) over an entire cross-section (y,z). It also says that the volume of water that is transported downwind shall be compensated by the same transport in the upwind direction. The upwind transport may develop underneath the downwind transport (vertically sheared) or to the side (laterally sheared). As we will see, the spatial response of the wind-driven flow throughout a cross-section, be it vertically or laterally sheared, will depend on its lateral bathymetry. The beauty of this approach is that it applies to any arbitrary lateral variations in bathymetry $H(y)$.

We can assess the flow response to wind forcing over a given cross-section by applying condition 6.15 to equation 6.8 and obtain a third relationship between slope and wind stress:

$$\frac{\partial\eta}{\partial x} = \frac{3}{2}\frac{\tau_s}{\rho_0 g}\left[\int_0^B H^2 dy \Big/ \int_0^B H^3 dy\right], \tag{6.16}$$

which is similar to and consistent with equation 6.10. The integrals in equation 6.16 represent scalars that can be determined numerically (e.g., with Matlab) for any depth distribution $H(y)$. This expression for water-level slope is then used in equation 6.8 to calculate and draw the wind-driven flow structure $[u(y,z)]$ over any cross-section $H(y)$. Keep in mind that the wind-driven flow (equation 6.8) has slope-driven parabolic profiles and stress-induced linear profiles. When combined, the resulting profiles are revealing. Over a triangular cross-section (as in Figure 6.3a), the wind-driven flow is downwind over the shallow portions of the cross-section and upwind in the channel. Upwind flow reaches all the way to the surface and develops over depths $H(y) > H_0/2$, where H_0 is the maximum depth.

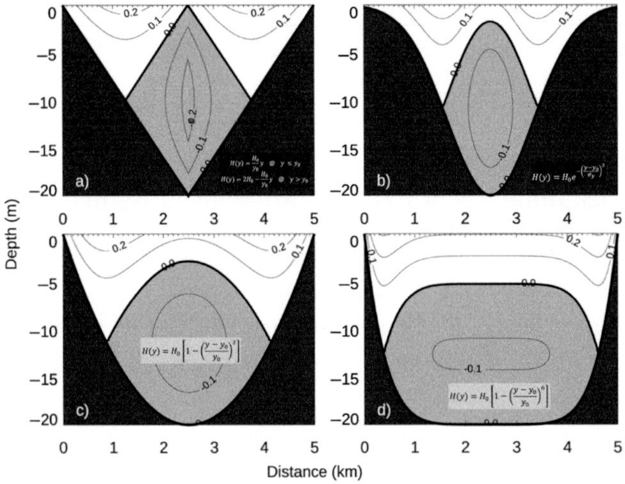

Figure 6.3 Wind-driven flow (eq. 6.8) using equation 6.16 to calculate the water-level slope. The bathymetric profile $H(y)$ is indicated by the equation on each panel. In those equations, H_0 is the maximum depth, y_0 is the position for H_0. The view on the figures can be either into or out of the semienclosed basin. Downwind flows are represented by contours (labelled in m/s) with white background. Upwind flows are shaded in gray. The zero isotach is the thickest contour. The cross-section width B is 5,000 m. Results are shown for τ_s of 0.05 Pa and A_z of 0.001 m^2/s.

When the bottom slope is less sharp, as in a Gaussian bathymetry (Figure 6.3b), the downwind flow still preferentially develops over shallow portions of the cross-section and upwind flow appears in the channel. The upwind flow reaches close to the surface but remains below it. As the bathymetric slope relaxes more over the middle part of the section (Figure 6.3c and d), compared to the previous two cases, the wind-driven exchange flow becomes increasingly sheared in the vertical direction. The diagrams illustrated in this figure can be reproduced with the Matlab script provided.

Perhaps the most complete (or general) way to relate the water-level slope to the wind stress is by assuming a non-zero net transport T (in m^3/s) through a cross-section:

$$\int_0^B \int_{-H}^0 u(y,z)dz\,dy = T. \tag{6.17}$$

As a reminder, this is the fourth way of trying to establish such a relationship, with the other three being given by equations 6.9, 6.12, and 6.15. Similar to condition 6.12, the transport T may or may not be related to wind forcing. It is challenging to determine whether the transport of water and the wind stress are related unless we have a robust relationship between them, which is still unknown. Applying equation 6.17 to the solution resulting from the balance between pressure gradient and friction (equation 6.8), we obtain such linkage between slope and wind stress:

$$\frac{\partial \eta}{\partial x} = \frac{3}{2}\frac{\tau_s}{\rho_0 g}\left[\frac{\int_0^B H^2 dy}{\int_0^B H^3 dy}\right] - \frac{3}{g}\frac{\mathbf{A}_z}{\int_0^B H^3 dy} T. \tag{6.18}$$

Equation 6.18, as well as equations 6.10, 6.13, and 6.16, provides a scalar (constant) value of $\partial\eta/\partial x$ that typically falls between 1×10^{-7} and 1×10^{-6}. This is the expected range of water-level slopes in semienclosed basins. The value obtained from equation 6.18 can be inserted in equation 6.8 to determine the modifications of T on the wind-driven flows that are portrayed in Figure 6.3. Recall that Figure 6.3 results are for zero net transport, that is, $T = 0$. Alterations by T will depend on the mean value and sign of the transport velocity T/A_c, where A_c is the cross-sectional area.

Taking the Gaussian bathymetry in Figure 6.3 and prescribing four different values of T offers instructive distributions. A transport of T = 2,000 m^3/s (in the same direction as the wind stress, Figure 6.4a) over the cross-section of 42,400 m^2 is equivalent to a cross-sectional mean flow of 0.047 m/s. This transport enhances the downwind flow and hinders the upwind flow in the channel. Consequently, the interface between outflow and inflow (zero isotach – although *isotach* strictly refers to wind speed, we are taking the liberty of using the same term for current

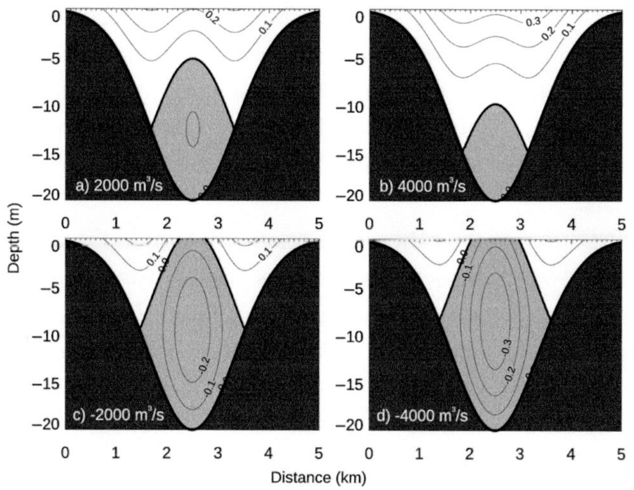

Figure 6.4 Wind-driven flow modified by remote transport T for a Gaussian bathymetry. The bathymetry is the same as in Figure 6.3b, where $T = 0$. The view on the figures can be either into or out of the semienclosed basin. Downwind flows are represented by contours with white background. Upwind flows are shaded in gray. The zero isotach is the thickest contour. Each panel displays flows with different magnitude and direction, and values for the prescribed transport T.

speed) migrates downward in the water column. Doubling the transport to 4,000 m^3/s drastically reduces upwind flow (Figure 6.4b) while the downwind flow dominates the cross-section. Further increases to T eventually transform the flow distribution into resembling an open-channel flow, like a river.

Transports in the opposite direction to the wind stress (Figure 6.4c and d) enhance the upwind flow while weakening the downwind flow. In this case, the upwind flow occupies most of the deepest portion of the cross-section. The zero isotachs become much steeper than with zero T so that they intersect the surface. Thus, the exchange flow in the semienclosed basin becomes laterally sheared instead of vertically sheared. As with the downwind transports, increasing upwind T should eventually cause upwind flows reminiscent of open-channel flow. This extreme situation can only appear in reality if the wind-driven flow is negligible compared to the upwind flow connected to T.

6.4 Comparison between Theory and Observations

This section presents two examples in which the theoretical results provided by equation 6.8 are compared to observed wind-driven flow fields across a section. In both instances, the cross-section was measured repeatedly during one full day with a towed acoustic Doppler current profiler (ADCP). Tidal and non-tidal signals were separated as in Appendix 4.1. In both cases the non-tidal flow was dominated by wind forcing. Because of such wind action, data collection was close to the limit of safety operations at sea. This is one of the main reasons there are few observational examples with enough spatial resolution of wind-driven circulation in semienclosed basins. "Enough" spatial resolution means that the data are able to resolve the flow structure over shoals, in the channel, and over transitions from channel to shoals.

The first illustration of the efficacy of relatively simple theoretical expressions for wind-driven flows in semienclosed basins comes from a coastal lagoon. The lagoon, Guaymas Bay, lies on the eastern coast of the Gulf of California and can be forced by northwesterly winds ~10 m/s in winter. Two consecutive-day winter experiments collected data during homogeneous conditions of water density. Residual flow distributions were consistent with theoretical results (Figure 6.5). Downwind flow occupied shoals and upwind flow appeared in the channel. In one of the experiments, winds were stronger than the other survey and upwind flows even reached close to the surface. Theoretical flows qualitatively resembled those observed. Differences in the details can be attributed to simplified dynamics in which the eddy viscosity is constant throughout the section, and to morphologic variability in the basin.

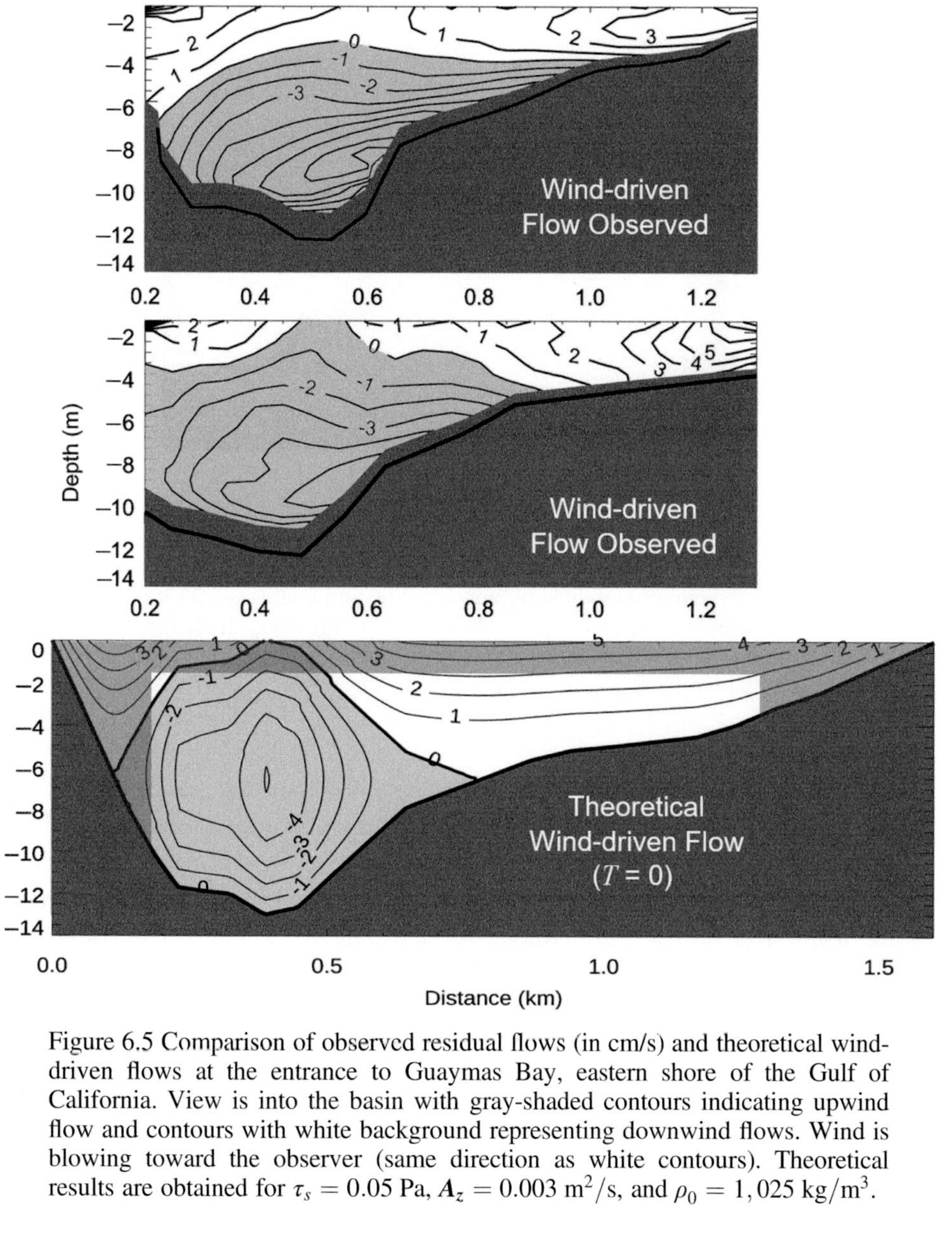

Figure 6.5 Comparison of observed residual flows (in cm/s) and theoretical wind-driven flows at the entrance to Guaymas Bay, eastern shore of the Gulf of California. View is into the basin with gray-shaded contours indicating upwind flow and contours with white background representing downwind flows. Wind is blowing toward the observer (same direction as white contours). Theoretical results are obtained for $\tau_s = 0.05$ Pa, $\boldsymbol{A}_z = 0.003\ \mathrm{m^2/s}$, and $\rho_0 = 1{,}025\ \mathrm{kg/m^3}$.

The second example displays data obtained in a bay of central Chile. Concepcion Bay is much wider and deeper than the lagoon of the previous example. The day-long survey occurred during stratified water columns with a well-defined pycnocline base at a depth of ~10 m. The observed residual flow had a downwind surface layer constrained by the pycnocline (Figure 6.6). The compensating upwind flow occupied most of the water column but had a

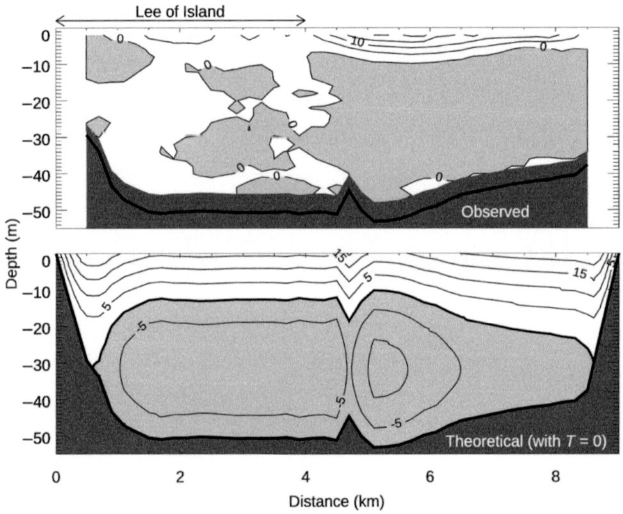

Figure 6.6 Comparison of observed residual flows (in cm/s) and theoretical wind-driven flows at the entrance to Concepcion Bay, central Chile. View is seaward with gray-shaded contours indicating upwind flow and contours with white background representing downwind flows. Wind is blowing away from the observer (same direction as white contours), representing upwelling conditions. Theoretical results are obtained for $\tau_s = 0.05$ Pa, $A_z = 0.003\ \mathrm{m}^2/\mathrm{s}$, and $\rho_0 = 1,025\ \mathrm{kg/m}^3$.

magnitude that was more than five times smaller than the strongest surface downwind flow. Although the theoretical expectation displays downwind surface flow and upwind flow underneath, the theoretical surface flow is much thicker than that observed. This is attributed to the pycnocline presence, which is disregarded by the theoretical results. Furthermore, the left half of the transect for observed flows lies on the lee effects of an island that disrupts the wind-driven flow. Again, the main discrepancy between observations and theoretical results, on the right half of the transect (Figure 6.6), is likely the assumption that the eddy viscosity is constant. Ongoing theoretical considerations are already studying spatially variable eddy viscosities pertinent for applications that are beyond this introductory treatment.

6.5 Take-Home Message

Wind-driven flows can be vertically sheared or laterally sheared, depending on a bathymetric cross-section. The response of a semienclosed basin to wind forcing can be represented with relatively simple dynamics consisting of pressure gradient from surface slopes and frictional effects. Quantitatively, the wind-driven flow distribution throughout a cross-section can be described with equation 6.8, which has a water-level slope contribution and a wind stress contribution. A relationship between water slope and wind stress provides the information to draw such wind-driven flow. Four different possibilities have been discussed in this chapter: (i) zero transport per unit width (for one profile); (ii) net transport in one profile; (iii) zero transport throughout a cross-section; and (iv) net transport throughout a cross-section.

Additional Sources

Winant, C.D. (2004) Three-dimensional wind-driven flow in an elongated, rotating basin. *J. Phys. Ocean.* 34(2): 462–476.

7

Flows Driven by Density Gradients

This chapter describes residual flows driven exclusively by density gradients. The previous chapter, in contrast, presented residual flows forced only by winds and Chapter 5 treated residual flows from tides. This chapter begins with a qualitative description of the flow arising from density gradients established by freshwater input to a semienclosed basin, that is, the *density-driven exchange flow* or *gravitational circulation*. It then goes into a dynamical description, grounded on fundamental physics. The dynamical description first considers lateral homogeneity and then allows for lateral variations caused by Earth's rotation. It then considers lateral changes in bathymetry and the relative contribution of frictional effects versus Earth's rotation effects. The chapter concludes by exploring the influence of basin width on gravitational circulation. Along-basin variations in residual flow are considered in Chapter 8.

7.1 Qualitative Description

Density-driven flows are the epitome of the interaction between riverine and oceanic waters in a semienclosed basin. Qualitatively, disregarding tidal effects for the moment, different water densities arise throughout a semienclosed basin when fresh waters enter the basin. Different water densities occur because fresh waters are lighter than ocean waters. The density gradient, thus established, responds to the influence of gravity by driving a long-term (days to weeks to months) propensity of the buoyant waters to flow toward the ocean near the surface. This surface outflow is compensated by a residual inflow of oceanic waters underneath. The result is that the baseline state of the freshwater-influenced basin is to exhibit outflow at the surface and inflow underneath. This type of long-term circulation is referred to as gravitational circulation or density-driven exchange flow. It is sometimes called estuarine circulation, but the estuarine circulation actually has other components related to tidal and wind forcing, and to the interaction of

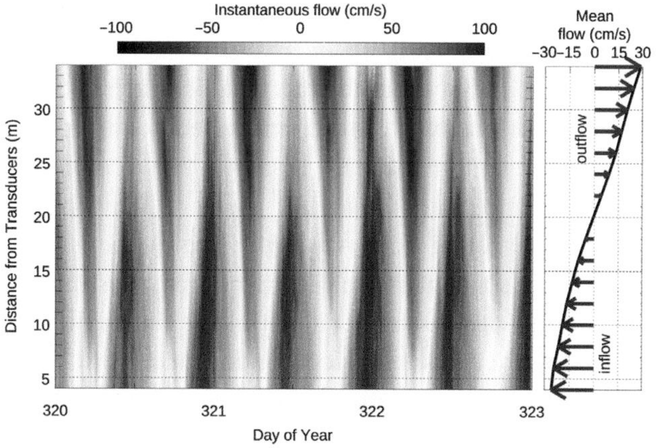

Figure 7.1 Contours of instantaneous along-basin velocity profiles obtained with an ADCP over a sill in a Chilean fjord over three d. To the right of the contours is a profile of the three-day average or residual circulation. The vertical axis displays distance from the ADCP transducers, themselves about 5 m from the bottom. The figure shows that the ebbs (blue shades) are stronger than the floods (red shades) at the surface, and that floods are strongest toward the bottom. The mean flow profile displays the typical gravitational circulation.

forcings. Hereafter, we will refer to the density-driven circulation, consisting of outflow at the surface and inflow underneath, as gravitational circulation.

In most tidal basins influenced by freshwater, the gravitational circulation will be observed only after averaging out the tidal variations. Instantaneous velocity profiles will exhibit vertical and temporal structures consistent with those illustrated in Figure 7.1. Ebb periods will show strongest seaward flows at the surface, furthest away from bottom frictional effects. Flood periods may exhibit strongest inflows underneath the surface (as in Figure 7.1) as the pressure gradient from the tides competes with the pressure gradient from the river. Instantaneous velocity profiles in Figure 7.1 are influenced by tidal currents that mask the gravitational circulation. A temporal average of the profiles in Figure 7.1, by eye and by calculation (mean flow), exhibits a well-developed gravitational circulation.

7.2 Quantitative Description: Along-Basin Momentum

This section presents the derivation of the classic gravitational circulation profile [$u(z)$], starting from the dynamic balance (equations 2.40) along the basin (x):

$$\underbrace{\frac{\partial u}{\partial t}}_{local} \underbrace{+u\frac{\partial u}{\partial x}+v\frac{\partial u}{\partial y}+w\frac{\partial u}{\partial z}}_{advective} - \underbrace{fv}_{Coriolis} = \underbrace{-g\frac{\partial \eta}{\partial x} - \int_{-H}^{z} \frac{g}{\rho_0}\frac{\partial \rho}{\partial x}dz}_{pressure\ gradient} + \underbrace{\frac{\partial}{\partial z}\left[A_z\frac{\partial u}{\partial z}\right]}_{friction}, \quad (7.1)$$

where v is the flow component in the lateral direction y, w is the component in the vertical z direction (positive upward, i.e., negative down the water column), f is the *Coriolis* parameter, g is gravity's acceleration, η is water elevation, ρ is water density, ρ_0 is a reference density, A_z is the vertical eddy viscosity, and H is water column depth. This momentum balance assumes that vertical frictional effects dominate horizontal friction, that is, that $A_z L_x^2 / A_h H^2 \gg 1$, where L_x is the basin's length and A_h is the horizontal eddy viscosity (see Sections 2.6 and 6.2). The pressure gradient force has a contribution from the water-level slope $\partial\eta/\partial x$ caused by river input, and from the density gradient along the basin $\partial\rho/\partial x$. The slope contribution to the pressure gradient is depth-independent while the density gradient contribution increases with depth, even for a constant (depth-independent) $\partial\rho/\partial x$ because of the integral in the vertical direction z.

7.2.1 Density-Driven Mean Flow

If we take equation 7.1 and assume that the gravitational circulation u is linear and at steady state (no local or advective accelerations), with no effects of rotation, and that A_z is constant, we can rewrite the along-basin momentum balance as

$$g\frac{\partial\eta}{\partial x} + \int_{-H}^{z} \frac{g}{\rho_0}\frac{\partial\rho}{\partial x}dz = A_z\frac{\partial^2 u}{\partial z^2}, \tag{7.2}$$

which describes the classical balance between pressure gradient (on the left), being the driving force of the circulation, and stress divergence, or vertical friction. The dynamics of equation 7.2 are elegant in their simplicity. If we also take the slope and the density gradients as constant, which is reasonable, equation 7.2 becomes

$$\frac{\partial^2 u}{\partial z^2} = \frac{g}{A_z}\frac{\partial\eta}{\partial x} - \frac{g}{\rho_0 A_z}\frac{\partial\rho}{\partial x}z. \tag{7.3}$$

This equation still describes the balance between pressure gradient and friction. It represents an ordinary differential equation. We can solve the equation for u (the dependent variable) as a function of depth (z, the independent variable) by integrating it twice relative to z. If the procedure to arrive at the solution is unimportant, skip to equation 7.13. Nonetheless, the procedure is rewarding. Integrating equation 7.3 once, yields

$$\frac{\partial u}{\partial z} = \frac{g}{A_z}\frac{\partial\eta}{\partial x}z - \frac{g}{2\rho_0 A_z}\frac{\partial\rho}{\partial x}z^2 + c_1, \tag{7.4}$$

where c_1 is an integration constant to be obtained with prescribed boundary conditions. Integrating equation 7.4 to obtain $u(z)$:

$$u(z) = \frac{g}{2A_z}\frac{\partial \eta}{\partial x}z^2 - \frac{g}{6\rho_0 A_z}\frac{\partial \rho}{\partial x}z^3 + c_1 z + c_2. \tag{7.5}$$

As in Section 6.2, integration constants c_1 and c_2 are obtained from two boundary conditions. The first condition is that at the surface ($z = 0$), the vertical shear (or the vertical gradient) in the flow u is provided by a wind stress, that is,

$$A_z \frac{\partial u}{\partial z}[z = 0] = \frac{\tau_s}{\rho_0}. \tag{7.6}$$

Another condition at the surface could be that the vertical shear in velocity is zero, in which case all terms involving τ in the following equations would also be zero. Applying equations 7.6 to 7.4 yields

$$c_1 = \frac{\tau_s}{\rho_0 A_z}. \tag{7.7}$$

The second boundary condition can be prescribed as no flow ($u = 0$) at the bottom ($z = -H$), also known as a "no-slip" condition. Other conditions may also be applied at the bottom such as (i) no shear at the bottom ($\partial u/\partial z = 0$) or (ii) prescribed bottom stress. Applying the no-slip condition to equation 7.5, with 7.7, gives (analogous to equation 6.7 but now including a density gradient)

$$0 = \frac{gH^2}{2A_z}\frac{\partial \eta}{\partial x} + \frac{gH^3}{6\rho_0 A_z}\frac{\partial \rho}{\partial x} - \frac{\tau_s H}{\rho_0 A_z} + c_2. \tag{7.8}$$

Substituting equations 7.7 and 7.8 in 7.5 provides the solution (also analogous to equation 6.8):

$$u(z) = \frac{gH^2}{2A_z}\frac{\partial \boldsymbol{\eta}}{\partial \boldsymbol{x}}\left[\frac{z^2}{H^2} - 1\right] - \frac{gH^3}{6\rho_0 A_z}\frac{\partial \boldsymbol{\rho}}{\partial \boldsymbol{x}}\left[1 + \frac{z^3}{H^3}\right] + \frac{\boldsymbol{\tau_s} H}{\rho_0 A_z}\left[1 + \frac{z}{H}\right]. \tag{7.9}$$

This solution indicates that the gravitational circulation is proportional to depth (actually to depth cubed) and inversely proportional to eddy viscosity (friction). The solution requires prescription of the water-level slope, the horizontal density gradient and the wind stress (all in bold in equation 7.9). However, in order to find dynamic consistency among these parameters, we need to find a relationship among them. One way to find this relationship, as we did in Chapter 6, is to assume that the volume flux (or transport) per unit width associated with equation 7.9 is non-zero (as in equation 6.12) and equal to the river discharge per unit width R_w:

$$\int_{-H}^{0} u(z)\partial z = R_w. \tag{7.10}$$

That is, equation 7.10 says that the river transport per unit width (in m^2/s) provides a volume added to the basin. Applying equations 7.10 to 7.9 results in

$$\boldsymbol{R_w} = -\frac{gH^3}{3A_z}\frac{\partial \boldsymbol{\eta}}{\partial \boldsymbol{x}} - \frac{gH^4}{8\rho_0 A_z}\frac{\partial \boldsymbol{\rho}}{\partial \boldsymbol{x}} + \frac{H^2}{2\rho_0 A_z}\boldsymbol{\tau_s}, \tag{7.11}$$

which provides a relationship between the water-level slope with the density gradient, the river transport and the wind stress:

$$\frac{\partial \boldsymbol{\eta}}{\partial \boldsymbol{x}} = -\frac{3}{8}\frac{H}{\rho_0}\frac{\partial \boldsymbol{\rho}}{\partial \boldsymbol{x}} - \frac{3A_z}{gH^3}\boldsymbol{R_w} + \frac{3}{2}\frac{\boldsymbol{\tau_s}}{\rho_0 gH}. \tag{7.12}$$

This is a nice relationship that describes the density gradient and river transport having opposite spatial variability to the water-level slope. Equation 7.12 is equivalent to equation 6.13 for a homogeneous fluid ($\partial\rho/\partial x = 0$). Also, R_w is analogous to T_x in equation 6.13 in the sense that both represent non-zero transports per unit width. Equation 7.12 reveals that when river discharge increases, the water-level slope in the receiving basin will increase in the opposite direction to the river flow direction. Similarly, when the density gradient increases in the basin, the water-level slope will increase with the opposite sign. Moreover, the water-level slope will respond directly to the direction of the wind: when the wind stress is positive, for example, down the basin toward the entrance, the water-level slope will also be positive; that is, water will pile up toward the basin's entrance. Finally, inserting equation 7.12 into the solution (equation 7.9), we obtain, after some fun algebra,

$$u(z) = \underbrace{\frac{gH^3}{48\rho_0 \boldsymbol{A_z}}\frac{\partial \boldsymbol{\rho}}{\partial \boldsymbol{x}}\left[9\left(1-\frac{z^2}{H^2}\right) - 8\left(1+\frac{z^3}{H^3}\right)\right]}_{\text{Density-induced}} + \underbrace{\frac{3}{2}\frac{\boldsymbol{R_w}}{H}\left[1-\frac{z^2}{H^2}\right]}_{\text{River-induced}}$$
$$+ \underbrace{\frac{1}{4}\frac{\boldsymbol{\tau_s} H}{\rho_0 \boldsymbol{A_z}}\left[4\left(1+\frac{z}{H}\right) - 3\left(1-\frac{z^2}{H^2}\right)\right]}_{\text{Wind-induced}} \tag{7.13}$$

This equation has three contributions: density-, river-, and wind-induced flows, which are assumed to be independent from each other. Each contribution has its free parameters (in bold in equations 7.11–7.13). The density-induced flow is inversely related to frictional effects ($\propto A_z^{-1}$) and highly sensitive to depth ($\propto H^3$). Its profile is described by a third-degree polynomial with two inflection points. The river-induced flow is sensitive to H^{-1} and depicts a parabolic profile. The wind-driven component also has a parabolic profile. The vertical integral of the solution $u(z)$ (equation 7.13) yields R_w, as indicated by equation 7.10. Each contribution to the solution is displayed in Figure 7.2, together with the sum of the three. Disregarding the wind-driven part represents the gravitational circulation.

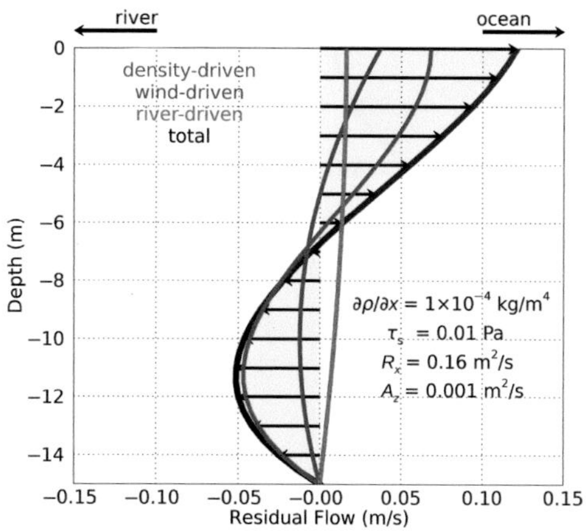

Figure 7.2 Velocity profile related to equation 7.13 resulting from the dynamic balance between pressure gradient and friction. Drawn are the separate contributions from wind stress, density gradients, and river discharge, with the corresponding free parameters of the solution. The vertical integral of $u(z)$ equals the contribution from river discharge, as per equation 7.10.

The circulation profile portrayed by equation 7.13 can be useful, despite its restrictive assumptions. In basins dominated by two-layer residual circulations, we can find the best fit between that theoretical expectation (equation 7.13) and the observed profile, as done in the exercise of Appendix 7.1. The exercise provides coding in Matlab for equation 7.13 and a least-squares-fit to the three free parameters in the equation, $[\partial\rho/\partial x/A_z]$, $[\tau_s/A_z]$, and R_w, to match to a given set of observations. More advanced solutions to the balance between pressure gradient and friction have adopted depth-dependent values of A_z (see Section 7.2.3).

Alternatively, to equation 7.13, if we take as bottom boundary condition $\partial u/\partial z = 0$ at $z = -H$, instead of the no-slip condition, the density-induced component becomes

$$u(z) = \frac{gH^3}{24\rho_0 A_z}\frac{\partial\rho}{\partial x}\left[1 - 4\frac{z^3}{H^3} - 6\frac{z^2}{H^2}\right], \tag{7.14}$$

which provides a symmetric profile around mid-depth (Figure 7.3a). This boundary condition can also be used to solve for the velocity profile in terms of the water-level slope, instead of the density gradient. In that case, equation 7.12 would be modified to express the density gradient in terms of the slope and then insert it into equation 7.9 to obtain

$$u(z) = \frac{gH^2}{12\rho_0 A_z}\frac{\partial\eta}{\partial x}\left[1 - 4\frac{z^3}{H^3} - 6\frac{z^2}{H^2}\right]. \tag{7.15}$$

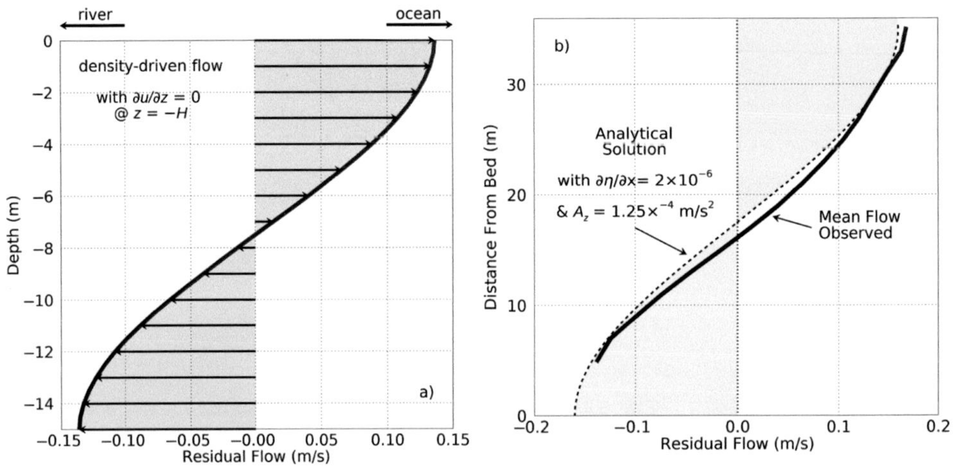

Figure 7.3 Mean velocity profiles with the condition of zero vertical shear at the bottom. (a) For density-driven flow exclusively (equation 7.14). (b) For slope-driven flow, dynamically consistent with density gradients (equation 7.15), compared to observed mean flow profile from 100 d of measurements in a Chilean fjord. Part of these observations are also portrayed in Figure 7.1.

This equation compares well to a 100-day mean velocity profile in the Inland Sea of Chile (Figure 7.3b), where water-level slope measurements were also available. In that comparison, the deployment-long mean value of $\partial\eta/\partial x$ was 2×10^{-6}, and the A_z was 1.25×10^{-4} m^2/s. Remember that equations 7.13 through 7.15 are obtained from a simplified momentum balance between pressure gradient and friction. In most fjord areas, however, frictional effects will be negligible compared to, say, advective momentum flux, and equation 7.15 will misrepresent the velocity profile. Figure 7.3b is one exception, perhaps because of increased local mixing.

7.2.2 Mean Salinity Profile

The gravitational circulation part of equation 7.13 (i.e., considering the density-driven and river-driven contributions while neglecting the wind-driven role) can also be expressed in terms of the horizontal salinity gradient $\partial S/\partial x$, remembering that

$$\frac{1}{\rho_0}\frac{\partial \rho}{\partial x} = \beta \frac{\partial S}{\partial x}, \tag{7.16}$$

where β is the coefficient of haline (or saline) contraction with values between 7.4×10^{-4} and 8.1×10^{-4} kg/g (see equation 2.28 and Figure 2.5b). With equation 7.16, the gravitational circulation of equation 7.13 becomes an equivalent:

$$u(z) = \frac{g\beta H^3}{48A_z}\frac{\partial S}{\partial x}\left[9\left(1-\frac{z^2}{H^2}\right)-8\left(1+\frac{z^3}{H^3}\right)\right]+\frac{3}{2}\frac{R_x}{H}\left[1-\frac{z^2}{H^2}\right]. \tag{7.17}$$

With this representation, we can also portray a theoretical salinity profile, $S(z)$, obtained from a simplified advection–diffusion salt balance (equation 2.21) that considers only along-basin advection of salinity and vertical mixing:

$$\underbrace{u\frac{\partial S}{\partial x}}_{advection} = \underbrace{K_z\frac{\partial^2 S}{\partial z^2}}_{vertical\ mixing}, \tag{7.18}$$

where K_z is the kinematic eddy diffusivity of salt in the vertical direction. This kinematic diffusivity has the same units as the kinematic eddy viscosity (m^2/s) and describes how fast and over what length scales salt is being redistributed vertically by turbulent processes. Values of K_z are typically larger than A_z because mass (salt) is more readily exchanged than momentum. The ratio A_z/K_z is known as the *turbulent Schmidt number* (Sc_t) and tends to range between 0.2 and 0.5, although it can sometimes be assumed to be 1.

The salt balance in equation 7.18 assumes steady state ($\partial S/\partial t = 0$), negligible lateral and vertical advection compared to along-basin advection. This means that $U_xL_y/U_yL_x \gg 1$, and $U_xH/U_zL_x \gg 1$, where U_x, U_y and U_z are typical mean velocities in the along-basin, lateral, and vertical directions, respectively; while L_x, L_y, and H are typical length scales in the corresponding directions. The last assumption is that vertical mixing of salt is much greater than the horizontal (both along and across the basin), that is, that $K_zL_x^2/K_hH^2 \gg 1$ and $K_zL_y^2/K_hH^2 \gg 1$, where K_h is the horizontal diffusivity of salt.

Equation 7.18 can also be solved by double integration to obtain the dependent variable S as a function of the independent variable z. The solution assumes that $\partial S/\partial x$, K_z, and A_z are constant and that the velocity vertical profile is given by equation 7.17. After integrating equation 7.18, we prescribe boundary conditions (a): no vertical S gradient at the bottom, that is, $\partial S/\partial z = 0$ at $z = -H$, and (b): that the depth-mean S is S_0, that is, $1/H\int_{-H}^{0} S\,dz = S_0$. Applying boundary conditions, after challenging algebra, we obtain

$$S(z)= \mathbf{S_0}+\frac{H^2}{\mathbf{K_z}}\frac{\partial \mathbf{S}}{\partial \mathbf{x}}\left[\frac{g\beta H^3}{48\mathbf{A_z}}\frac{\partial \mathbf{S}}{\partial \mathbf{x}}\left(-\frac{1}{12}+\frac{1}{2}\frac{z^2}{H^2}-\frac{3}{4}\frac{z^4}{H^4}-\frac{2}{5}\frac{z^5}{H^5}\right)+\frac{\mathbf{R_x}}{H}\left(-\frac{7}{120}+\frac{1}{4}\frac{z^2}{H^2}-\frac{1}{8}\frac{z^4}{H^4}\right)\right]. \tag{7.19}$$

The mean salinity profile is thus depicted by a fifth-degree polynomial. It has contributions from the salinity gradient and river transport per unit basin's width. For a given depth H, there are five free parameters in this expression (in bold) that allow drawing of a profile: the horizontal salinity gradient, the vertical eddy viscosity, the vertical eddy diffusivity of salt, the depth-mean salinity, and the river discharge. Associated with Appendix 7.1, we provide a Matlab program that compares equation 7.19 to an observed mean salinity profile.

Vertical stratification portrayed by the salinity profile in equation 7.19 is sensitive to the eddy coefficients A_z and K_z, but especially to water column depth H (Figure 7.4a). In fact, the salinity contrast from surface to bottom is inversely proportional to the product $K_z A_z$. Small increases in the eddy coefficient (say, from $10^{-3.2}$ to 10^{-3}) may cause the profile to move from strongly stratified to weakly stratified. However, since the profile is sensitive to $H^2 \times H^3 = H^5$, a depth increase from 10 to 12 m causes a noticeable increase in stratification (Figure 7.4b and c) for the same value of A_z.

7.2.3 Depth-Dependent Eddy Viscosity

When using equations 7.13 and 7.19 we should remember the restrictions of the solutions. The main constraint is perhaps the assumption of constant eddy coefficients. This limitation allows the salinity to increase almost linearly within the water column, away from the boundaries where the vertical gradient tends to zero. This approach would be unsuitable in fjords, for example, where a distinct pycnocline indicates that the eddy coefficients change with depth. Sometimes, an observed mean salinity profile follows a similar distribution to equation 7.19, which would tell us that the assumptions made for that equation should be valid at the place and time of observations.

In the momentum balance given by equations 7.2 and 7.3, the eddy viscosity is constant. An alternative residual flow profile may be obtained by prescribing a depth-dependent A_z with a parabolic distribution that resembles observations in the field. The eddy viscosity may have the following depth-dependent distribution:

$$A_z = -\kappa u_* \left[\frac{z^2}{H} + \left(\frac{z_0}{H} + 1 \right) z \right], \tag{7.20}$$

where κ is the nondimensional *von Karman's constant* ($= 0.4$), u_* is a bottom friction velocity (in m/s, typically equal $\sqrt{\tau_b/\rho_0}$, τ_b being the bottom stress), z_0 is a bottom roughness length scale (in m), and z goes from the bottom $z = -H$ to the surface $z = 0$. A Matlab script is provided to generate a plot of this A_z profile. Using the expression for A_z in equation 7.20 and solving the analogous to equation 7.2 with boundary conditions of no flux at the surface, no-slip at the bottom, and vertically averaged flow given by the river velocity u_R, we get (after some work)

$$u(z) = \underbrace{\frac{gH^2}{2\rho_0 \kappa \boldsymbol{u}_*} \frac{\partial \boldsymbol{\rho}}{\partial \boldsymbol{x}} \left[\frac{1}{2c_i} \ln\left(\frac{z + H + z_0}{z_0} \right) - \left(1 + \frac{z}{H} \right) \right]}_{density-driven} + \underbrace{\frac{\boldsymbol{u_R}}{c_i} \ln\left(\frac{z + H + z_0}{z_0} \right)}_{river-driven}, \tag{7.21}$$

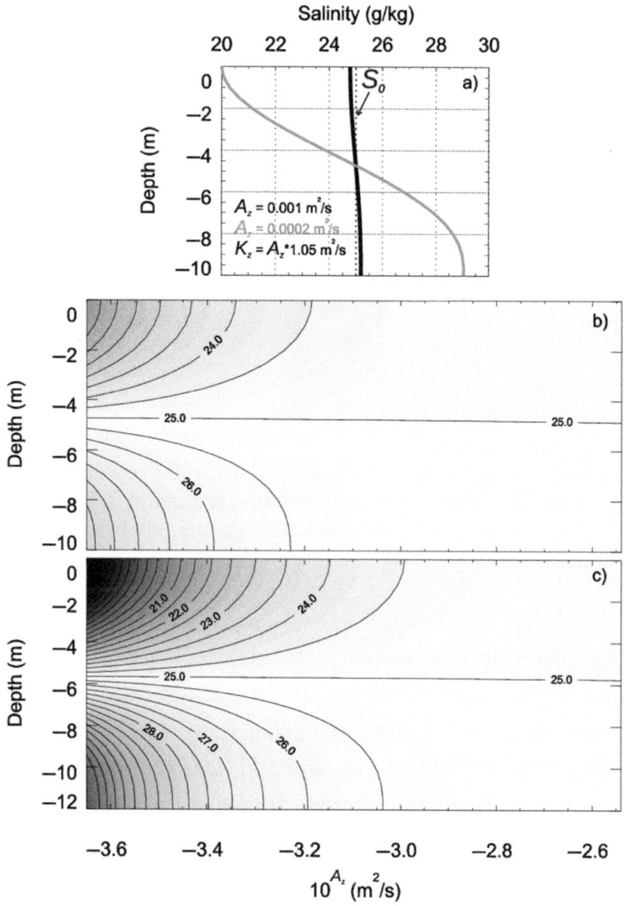

Figure 7.4 Salinity profiles for various eddy viscosities and depths. (a) Shows two salinity profiles around the mean, S_O, for the corresponding values of eddy viscosities. In both cases, the eddy diffusivity is greater than the eddy viscosity by 5% (Sc_t = 0.95). (b) Contours of salinity values as a function of eddy viscosity and depth for a water column H of 10 m. The abscissa shows values representing the logarithm, base 10, of the actual eddy viscosity. Diffusivities in the calculation remain 5% higher than the viscosity (Sc_t = 0.95). (c) is the same as (b) but for H = 12 m. Contour interval in (b) and (c) is 0.5 g/kg.

obtaining the constant of integration

$$c_i = \frac{z_0 + H}{H} \, ln \left(\frac{z_0 + H}{z_0} \right) - 1. \tag{7.22}$$

The same Matlab script (provided) that plots equation 7.20 plots equation 7.21 with 7.22. The velocity profile thus obtained exhibits almost a linear distribution with depth (Figure 7.5). This velocity profile reaches the inflection point from maximum

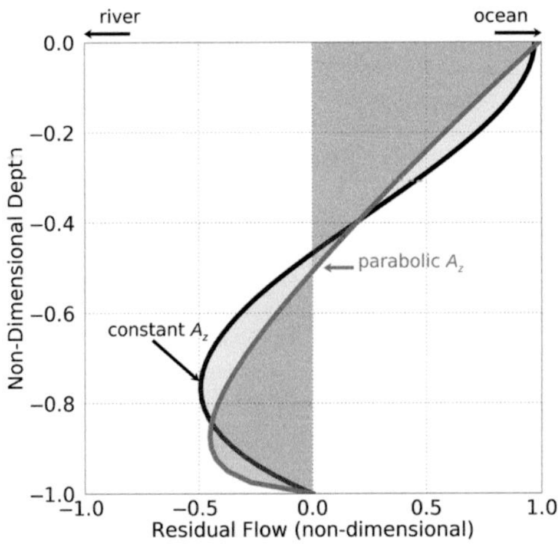

Figure 7.5 Comparison of theoretical expressions for gravitational circulation assuming constant (equation 7.13) and parabolic (equation 7.21) eddy viscosities. Qualitatively, both profiles indicate consistent results of outflow near the surface and inflow underneath.

inflow (negative values) to zero velocity, at a depth closer to the bottom than in the profile with constant A_z. In other words, the maximum inflow is reached deeper in the water column than with the case of constant A_z. Four free parameters in equation 7.21 are shown in bold (u_*, z_0, u_R, and $\partial\rho/\partial x$). Enhanced friction velocity will weaken the gravitational circulation. Increased roughness length scale will move the maximum inflow further away from the bottom. Augmented river flow will intensify the outflow portion of the profile. Finally, greater density gradients will strengthen exchange flows. As done in Appendix 7.1 for equation 7.13, equation 7.21 may be matched to observations by finding the optimal free parameters.

Presented throughout this section are the simplest along-basin momentum and salt balances. They establish the basic along-basin dynamical framework to study semienclosed basins driven by a pressure gradient. To complete this basic dynamical framework, we now turn our attention to the momentum balance across the basin.

7.3 Across-Basin Momentum

This section presents the simplest across-basin momentum balance that supports the gravitational circulation $[u(z)]$ explained in the previous section. Beginning with the dynamic balance across the basin (y), again taking $A_z L_x^2 / A_h H^2 \gg 1$ and $A_z L_y^2 / A_h H^2 \gg 1$ to neglect horizontal friction:

$$\underbrace{\frac{\partial v}{\partial t}}_{local} + \underbrace{u\frac{\partial v}{\partial x} + v\frac{\partial v}{\partial y} + w\frac{\partial v}{\partial z}}_{advective} + \underbrace{fu}_{Coriolis} = \underbrace{-\frac{1}{\rho_0}\frac{\partial p}{\partial y}}_{pressure\ gradient} + \underbrace{\frac{\partial}{\partial z}\left[A_z\frac{\partial v}{\partial z}\right]}_{friction} . \tag{7.23}$$

The simplest dynamical framework is maintained by assuming no lateral flows ($v = 0$), which transforms the across-basin momentum balance into (as all terms containing v vanish)

$$\underbrace{fu}_{Coriolis} = \underbrace{-\frac{1}{\rho_0}\frac{\partial p}{\partial y}}_{pressure\ gradient} , \tag{7.24}$$

known as the *geostrophic balance.* We can determine the validity of this approach in a semienclosed basin by scaling any of these two terms in relation to other possible forces, for example, friction. A relatively simple approach of carrying out an assessment of the validity of the geostrophic approximation involves the *level of no-motion*, which is the depth at which u in the gravitational circulation profile becomes zero. To validate the geostrophic approximation in the transverse direction, compare the observed slope of the interface of "no motion" across a basin to that expected from geostrophy (see Figure 7.6). The expected slope can be derived by looking at the geostrophic balance in the assumed homogeneous upper and lower layers that exchange between the basin and the adjacent coastal ocean.

The geostrophic balance in the lower layer (Figure 7.6) may be written as

$$fu_2 = -\frac{1}{\rho_2}\frac{\partial p_2}{\partial y}, \tag{7.25}$$

but p_2 may be written as $p_2 = \rho_1 g h_1 + \rho_2 g h_2$, which turns equation 7.25 into

$$fu_2 = -\frac{1}{\rho_2}\left[\rho_1 g\frac{\partial h_1}{\partial y} + \rho_2 g\frac{\partial h_2}{\partial y}\right]. \tag{7.26}$$

The geostrophic balance in the upper layer may be written in terms of the lateral surface slope $\partial\eta/\partial y$ as

$$fu_1 = -g\frac{\partial \eta}{\partial y}. \tag{7.27}$$

With equations 7.26 and 7.27, we seek to obtain an expression for the interface slope (slope of the level of no motion) (or $\partial h_2/\partial y$). Because

$$\frac{\partial \eta}{\partial y} = \frac{\partial}{\partial y}[h_1 + h_2], \tag{7.28}$$

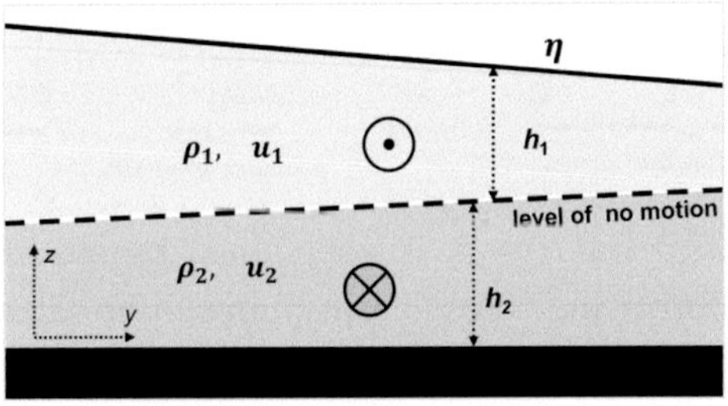

Figure 7.6 Diagram of the cross-basin distribution of two-layer exchange flow (looking into the basin in the northern hemisphere). The exchange is assumed to be uniform in speed and density throughout an upper and a lower layer. The *level of no motion* is the interface between inflow and outflow (zero isotach) and can also be regarded as the slope (in the vertical) of a front.

then, equation 7.27 may be rewritten as

$$-\frac{fu_1}{g} = \frac{\partial \eta}{\partial y} = \frac{\partial h_1}{\partial y} + \frac{\partial h_2}{\partial y}, \tag{7.29}$$

solving for $\partial h_1/\partial y$

$$\frac{\partial h_1}{\partial y} = -\frac{fu_1}{g} - \frac{\partial h_2}{\partial y}, \tag{7.30}$$

which also represents the geostrophic balance in the upper layer. Substituting equation 7.30 in 7.26 yields

$$fu_2 = -\frac{1}{\rho_2}\left[-\rho_1 g\left(\frac{fu_1}{g} + \frac{\partial h_2}{\partial y}\right) + \rho_2 g\frac{\partial h_2}{\partial y}\right]. \tag{7.31}$$

Finally, solving for $\partial h_2/\partial y$, that is, the slope of the interface

$$\frac{\partial h_2}{\partial y} = \frac{f(u_1\rho_1 - u_2\rho_2)}{g(\rho_2 - \rho_1)}, \tag{7.32}$$

is known as *Margules' equation*. The interface slope, which actually can be a frontal slope, depends then on the density and velocity contrast between upper and lower layers and the latitude (*Coriolis* parameter). Remember that this relationship is attained by assuming geostrophic dynamics in the cross-basin direction. This concept is borrowed from meteorology where it has been used to study atmospheric fronts.

Margules' relation can provide an initial tool to diagnose whether observed frontal features in a semienclosed basin are in geostrophic balance. This idea is explored with tidally averaged salinity data collected at the entrance to Chesapeake Bay (Figure 7.7), where the *Coriolis* parameter f equals 8.8×10^{-5} s^{-1}. The data

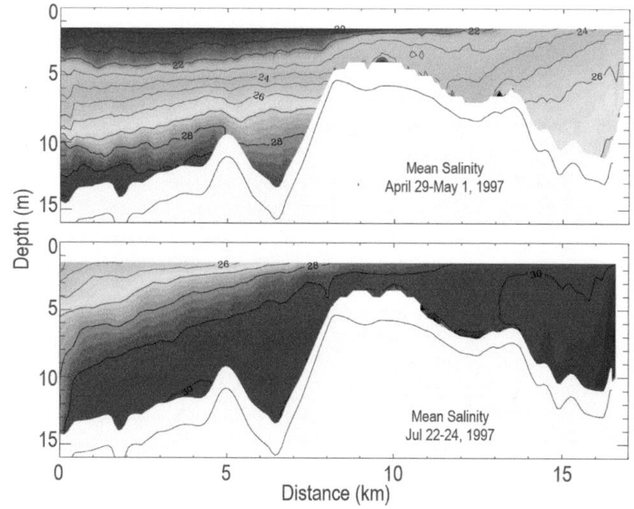

Figure 7.7 Mean salinity cross-sections at the entrance to Chesapeake Bay under different freshwater discharge conditions. The two sections illustrate distinct conditions for which the geostrophic approximation may or may not apply in the cross-basin direction. View is into Chesapeake Bay, from the ocean

were obtained over four consecutive tidal cycles towing an undulating conductivity-temperature-depth (CTD) recorder. Data were available in the spring, under the highest freshwater influence of the year, and under much weaker freshwater influence in the summer. For the spring salinity field, the isohaline of 24 g/kg could be taken as the interface (Figure 7.7, upper panel). Over the first 8 km of the transect, the depth of the isohaline changed from 7 to 5 m, that is, 2 m in 8 km, for a slope of 2.5×10^{-4}. Measurements indicated $u_1 = 0.10\ \mathrm{m/s}$ (mean outflow), $u_2 = -0.05\ \mathrm{m/s}$ (mean inflow), $\rho_1 = 1{,}017\ \mathrm{kg/m^3}$ and $\rho_2 = 1{,}022\ \mathrm{kg/m^3}$. Margules' equation predicted a slope of 2.7×10^{-4}, which was remarkably close to that observed. In the first 8 km of the transect, at least, the dynamics was approximately geostrophic. Between 10 and 17 km, the slope of the isohalines became steeper because of vertical mixing, indicating a departure from geostrophy.

The summer distribution displayed much higher salinities than in the spring (Figure 7.7, lower panel). The isohalines showed a steeper slope and the water column exhibited weaker vertical stratification throughout, relative to the spring. Taking the 28 g/kg isohaline as the interface, it shoaled 7 m (from 8 m at 0 km, to 1 m at 8 km) in 8 km, indicating a slope of 8.8×10^{-4}. Using observed values of $u_1 = 0.08\ \mathrm{m/s}$, $u_2 = -0.06\ \mathrm{m/s}$, $\rho_1 = 1{,}021\ \mathrm{kg/m^3}$ and $\rho_2 = 1{,}023\ \mathrm{kg/m^3}$, Margules' equation provided a slope of 6.4×10^{-4}. Although the difference between theoretical and observed slopes was larger for the summer example than for the one in the spring, it suggests that the geostrophic balance was a reasonable

Figure 7.8 Image of a buoyant discharge from the Elwha River entering Juan de Fuca Strait, in the state of Washington, USA. The discharge expands offshore to a distance R_i, where it forms a coastal current constrained by such a scale.

approximation to the cross-basin dynamics. The greater departure from geostrophy in the summer was evident by the increased slope in the isohalines, suggesting the influence of frictional effects as explored in Section 7.4.

Another useful application of geostrophy is to help determine the expansion limit, or expansion scale, of a buoyant discharge onto a semienclosed basin (e.g., Figure 7.8). When the discharge is affected by Earth's rotation, it will expand offshore and then be deflected by *Coriolis* acceleration along a boundary current. The width of the current R_i can be derived by scaling one form of the geostrophic balance:

$$fv = \frac{g}{\rho_0}\frac{\partial \rho}{\partial x}H. \tag{7.33}$$

Scaling the velocity v as $f R_i$, depth as H, the horizontal scale ∂x as R_i, and calling $g\partial\rho/\rho_0 = g'$ the reduced gravity (2–3 orders of magnitude $<g$), equation 7.33 scales as

$$f^2 R_i = \frac{g'H}{R_i}. \tag{7.34}$$

Solving for R_i, the scale of concern, we get

$$R_i = \frac{\sqrt{g'H}}{f}, \tag{7.35}$$

which is known as the *internal radius of deformation* or *internal Rossby radius*. This is a length scale that constrains a geostrophic balance in a buoyant discharge. In other words, it is the width of a buoyant discharge in geostrophic balance. In equation 7.35, g' is called *reduced gravity* because g is scaled by the ratio $\partial\rho/\rho_0$, which is typically 1,000 to 100 times smaller than g. Also, in equation 7.35, the numerator represents the phase speed of a long internal wave. In a way, equation 7.35 provides the scale over which the long internal wave (one could think of it as a buoyant front) is affected by Earth's rotation.

The basic dynamical framework built thus far is such that along the basin we have the balance between pressure gradient and friction, while across the basin we have a geostrophic balance. This is a framework that was proposed in the early studies of estuarine hydrodynamics and is consistent with the vertically sheared gravitational circulation comprised of outflow in a surface layer and inflow underneath. Through decades of research, we have learned that this framework represents a decent approximation to the dynamics but that some modifications provide improved understanding. One factor that certainly modifies the vertically sheared gravitational circulation is the variation of bathymetry across a basin.

7.4 Lateral Variations in Bathymetry

Transverse or lateral variations in bathymetry can decidedly modify the gravitational circulation. The same momentum balance as in equation 7.2, that is, pressure gradient driving the flow and friction counteracting it, can be solved over arbitrary lateral bathymetries $H(y)$. The solution is still equation 7.9 but the condition is now

$$\int_0^B \int_{-H}^0 u(z)\partial z dy = R, \tag{7.36}$$

where R is the river discharge (m^3/s). Applying this condition to equation 7.9 to obtain the water-level slope as a function of $\partial\rho/\partial x$, R, and τ_s results in

$$\frac{\partial\eta}{\partial x} = -\frac{3}{8\rho_0}\frac{\partial\rho}{\partial x}\frac{\int_0^B H^4 dy}{\int_0^B H^3 dy} - \frac{3A_z}{g\int_0^B H^3 dy}R + \frac{3}{2}\frac{\tau_s}{\rho_0 g}\frac{\int_0^B H^2 dy}{\int_0^B H^3 dy}. \tag{7.37}$$

Solutions for equation 7.9, inserting equation 7.37, are obtained for various bathymetries and $R = 0$ (Figure 7.9). It appears that when the flow "feels" the frictional changes caused by lateral variations in bathymetry (Figure 7.9a–c), the exchange flow is modified relative to flat-bottomed sections (Figure 7.9d). Instead of being bidirectionally vertically sheared, displaying outflow at the surface and inflow underneath, the exchange flow becomes laterally sheared. Outflow concentrates

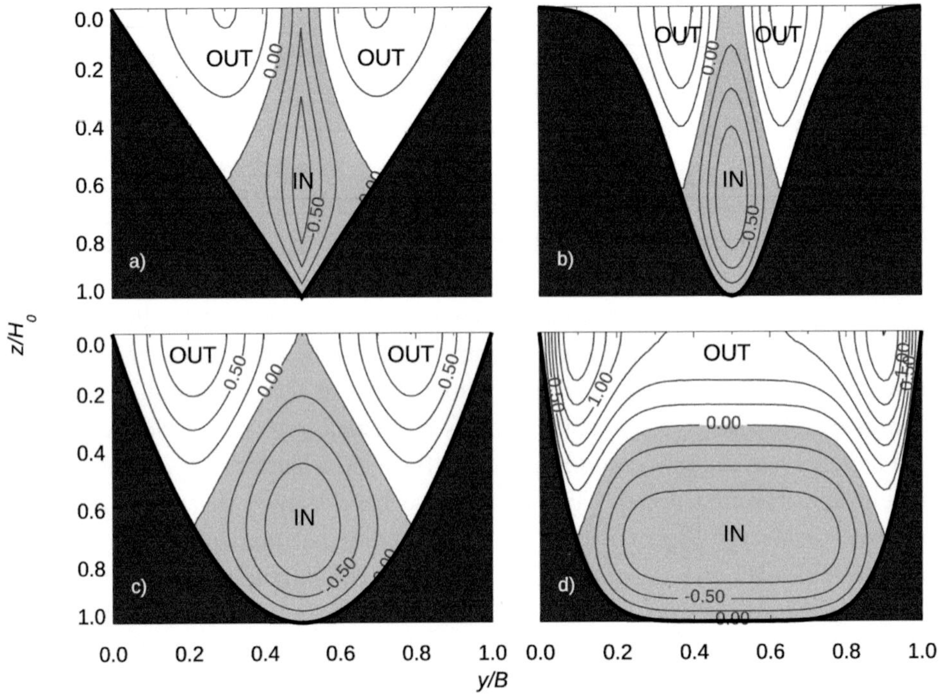

Figure 7.9 Density-driven circulation that develops over cross-sections with varied arbitrary bathymetries [$H(y)$]. Bathymetries are shaded in dark gray. View is into the basin. Light gray contours indicate inflow. As lateral variations in bathymetry become increasingly pronounced, exchange flows become more laterally sheared with inflow in channel and outflow over shoals. Over relatively flat bottoms (d), the exchange flow is vertically sheared.

over shoals and inflow appears in the channel. When bathymetric changes in the vicinity of the deepest portion of the cross-section are sufficiently steep, the exchange flow is laterally sheared.

Inclusion of R (i.e., $R \neq 0$ in equation 7.37) modifies the flow structure according to its direction, that is, seaward, or may also be prescribed landward (Figure 7.10). Volume flux in the direction of the surface outflow hinders inflow (Figure 7.10b and c). In this case, the inflow that reaches the surface with no discharge becomes detached from the surface. When R flows in the direction of inflow, outflow is hampered and the flow into the basin concentrates in the deepest portion of the cross-section, reaching the surface throughout a larger portion than with no discharge. These relatively simple solutions (provided in a Matlab script) may be comparable to observed flow distributions.

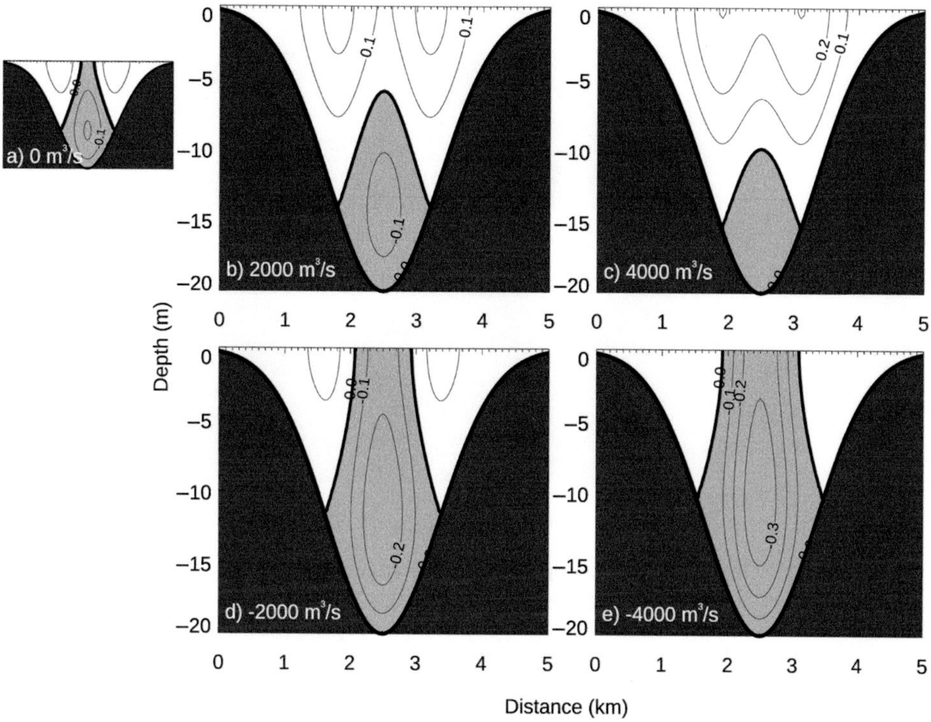

Figure 7.10 Density-driven circulation modified by river discharge R over an arbitrary cross-section. Same as Figure 7.9 but for different values of R, with the gravitational outflow (b and c) and with the gravitational inflow (d and e). Reference conditions ($R = 0$) are given in (a).

Observational examples abound of the flow structures illustrated in Figures 7.9 and 7.10. They are embodied here with two cross-sections obtained in the lower Chesapeake Bay (Figure 7.11). Right at the entrance to the bay (Figure 7.11a), the exchange flow is bidirectionally vertically sheared in the deep channel, between 2 and 4 km. In practically the rest of the section, the residual flow is seaward. In contrast, another section roughly 2 km landward from the previous transect displayed laterally sheared exchange flows with inflow in channels extending all the way to the surface. The flow field in both cross-sections, however, showed influence of Earth's rotation as it displayed slight deflection to its right. A question that arises is, Why is the flow structure so different despite the proximity of the transects? One plausible explanation involves two parts that alter the dynamical framework established in Sections 7.2 and 7.3. First, that the cross-basin dynamics depart from geostrophy and may have effects from friction. Second, that the along-basin dynamics have influence from *Coriolis* accelerations.

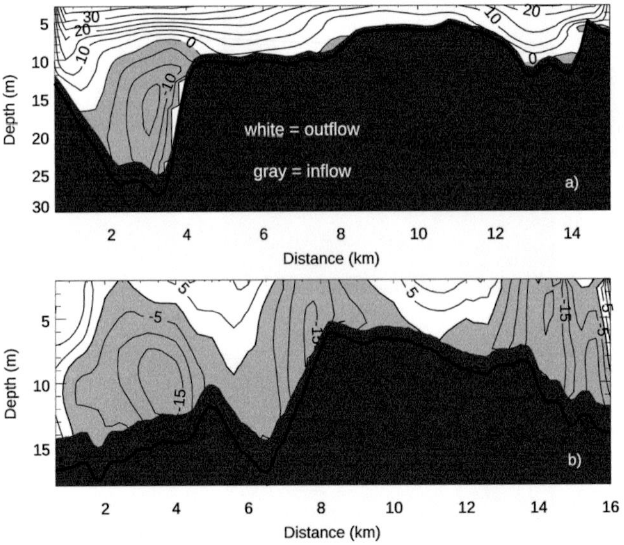

Figure 7.11 Observed residual flows in the lower Chesapeake Bay after tidal-cycle measurements along each transect. Looking into the estuary with gray shaded contours describing residual inflows. Black line represents the bathymetry and the dark gray shades mask the bathymetry and acoustic side-lobe effects near the bottom. (a) Right at the entrance and (b) nearly 2 km landward from (a). The transect in (b) is the same as in Figure 7.7 in the summer.

To study these possible explanations, the dynamical framework to study density-driven flows may be modified as follows:

$$\text{along-basin}: \underbrace{-fv}_{Coriolis} = \underbrace{-g\frac{\partial \eta}{\partial x} + \frac{g}{\rho}\int_{-H}^{0} \frac{\partial \rho}{\partial x} dz}_{pressure\ gradient} + \underbrace{A_z \frac{\partial^2 u}{\partial z^2}}_{friction} \tag{7.38}$$

$$\text{along-basin}: \underbrace{fu}_{Coriolis} = \underbrace{-g\frac{\partial \eta}{\partial y} + \frac{g}{\rho}\int_{-H}^{0} \frac{\partial \rho}{\partial y} dz}_{pressure\ gradient} + \underbrace{A_z \frac{\partial^2 v}{\partial z^2}}_{friction}. \tag{7.39}$$

The driving force is the pressure gradient, which is balanced by either *Coriolis* acceleration or friction, or both. In this framework, the dynamics are symmetric for both momentum components. Friction competes with *Coriolis* to determine which term will balance the pressure gradient. The competition between these two agents, friction and *Coriolis*, can be assessed with a nondimensional number obtained from scaling both forces:

$$\frac{friction}{Coriolis} = \frac{A_z}{fH^2} = E_k, \tag{7.40}$$

also known as the vertical *Ekman number*. This number compares the depth of frictional influence A_z/fH, analogous to an Ekman layer (in meters), to the water column depth. In a way, the Ekman number E_k represents a dynamical depth of a basin. A small E_k (< 0.01) indicates that the depth of frictional influence is negligible compared to the water column depth, that is, a *dynamically deep* basin. Similarly, a large E_k (> 1) implies that friction influences the entire water column, that is, a *dynamically shallow* basin. Intermediate E_k denotes that both friction and *Coriolis* affect the dynamics by balancing the driving force (pressure gradient).

Equations 7.38 and 7.39 are a pair of partial differential equations that can be solved analytically for the velocity (u, v) as a function of distance across the basin y and depth z. The solution applies to any arbitrary bathymetry $H(y)$ and is formulated in terms of a velocity that has a real part (along the basin) and an imaginary part (across the basin) $\mathcal{U} = u + iv$:

$$\mathcal{U}(y,z) = -i\left\{\underbrace{\frac{gN}{f}\left[1 - \frac{\cosh\delta z}{\cosh\delta H}\right]}_{slope-driven} + \underbrace{\frac{D}{f\alpha}\left[(e^{\delta z} - \delta z) - (e^{-\delta H} + \delta H)\frac{\cosh\delta z}{\cosh\delta H}\right]}_{density-driven}\right\}. \tag{7.41}$$

In solution 7.41, z is the vertical distance from the surface (negative downward), N is the water-level slope ($N = \partial\eta/\partial x + i\partial\eta/\partial y$), D is the density gradient ($D = g/\rho_0[\partial\rho/\partial x + i\partial\rho/\partial y]$), and $\delta = (1+i)/D_E$, with $D_E = \sqrt{2A_z/f}$ being the depth of the Ekman layer. In a way, δ is the inverse of the depth of bottom frictional influence. The free parameters in this solution are A_z, N, and D. With these three parameters, we obtain a velocity profile $\mathcal{U}(z)$ at each location y of a cross-section of width B for prescribed values of $H(y)$ and f.

As done in equations 7.9 through 7.11, we need a dynamically consistent relationship between N and D (water slope and density gradient). This is obtained by specifying that the net transport through the cross-section is provided by a river discharge $R = R_x + iR_y$, that is,

$$\int_0^B \int_{-H}^0 \mathcal{U}(y,z)\partial z dy = R. \tag{7.42}$$

The dynamically consistent relationship that determines the density gradient by prescribing N is

$$D = \frac{if\delta^2 R - \delta g \int_0^B N(y)[\tan h(\delta H) - \delta H]dy}{\int_0^B \left[\tan h(\delta H)(e^{-\delta H} + \delta H) - 1 + e^{-\delta H} - \delta^2 H^2/2\right]dy}. \tag{7.43}$$

This solution (equations 7.41 and 7.43) is portrayed in Figure 7.11 for the northern hemisphere and $R = 0$ (only density-driven). Solutions are drawn under prescribed $H(y)$, f and $N(y)$, and for different eddy viscosities, ultimately for different Ekman numbers. Prescription of $N(y)$ should have a form like

$$N(y) = N_0\left(1 + ie^{-\varsigma^2 y^2}\right) \tag{7.44}$$

to produce reasonable results with expected lateral variations produced by *Coriolis* accelerations. In this expression, ς is the rate of decay of the water-level slope $\partial\eta/\partial y$. It could be regarded as the inverse of the internal radius of deformation, that is, $\varsigma = R_i^{-1}$.

In Figure 7.12, the Ekman numbers are calculated with the maximum value of $H(y)$ and therefore may be slightly undervalued. It is evident that for low E_k (1×10^{-4}) the exchange flow is vertically sheared and influenced by *Coriolis* accelerations. Lateral variations in along-basin flow (contours in Figure 7.12) are caused by Earth's rotation as frictional effects are negligible (dynamically deep basin). This flow structure is similar to that observed in the Chesapeake Channel exchange flows in Figure 7.11a. In Figure 7.12, lateral flows for low E_k are linked to along-basin flows through rotation, describing a counterclockwise circulation (looking into the basin). As E_k grows, exchange flows become increasingly influenced by bathymetry as inflows remain in the channel and tend to occupy the entire water column. Outflows tend to be restricted to the shoals and form two branches separated by the inflow. This situation emulates the observed structure in Figure 7.11b. When *Coriolis* accelerations become irrelevant (dynamically shallow basins, $E_k \sim 1$), the outflows over shoals become symmetric and the inflow occupies the entire water column, that is, the exchange flows become symmetric about the deepest part of the channel. Furthermore, lateral flows become weaker as E_k increases.

Similar solutions to those portrayed in Figure 7.12 have been proposed for conditions that differ from equation 7.42. A commonly used condition is that the integral of lateral flows is zero at each lateral velocity profile. This condition may be attractive but produces flow structures that depart from observations. Although results similar to Figure 7.12 have compared favorably with observations in several estuaries, a more comprehensive condition than equation 7.42 is still needed.

A question that arises from the exchange flows portrayal in Figure 7.12 is, What role does the width of the basin B play in the structure of the flows? Figure 7.12 displays results for different E_k but the same B. One way to explore width effects on the exchange structure is to cast solution 7.41 in terms of a nondimensional scale. Comparing B to the internal Rossby radius R_i (e.g., equation 7.35) we obtain the *Kelvin number*:

$$K_e = B/R_i. \tag{7.45}$$

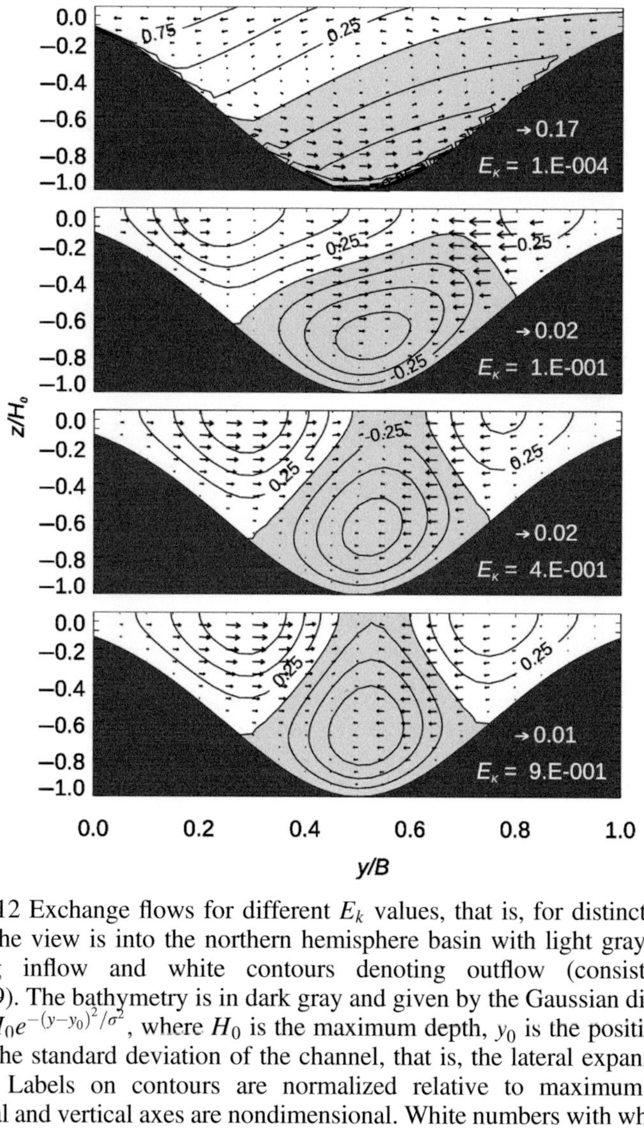

Figure 7.12 Exchange flows for different E_k values, that is, for distinct dynamic depths. The view is into the northern hemisphere basin with light gray contours indicating inflow and white contours denoting outflow (consistent with Figure 7.9). The bathymetry is in dark gray and given by the Gaussian distribution $H(y) = H_0 e^{-(y-y_0)^2/\sigma^2}$, where H_0 is the maximum depth, y_0 is the position of H_0, and σ is the standard deviation of the channel, that is, the lateral expansion of H from y_0. Labels on contours are normalized relative to maximum outflow. Horizontal and vertical axes are nondimensional. White numbers with white lateral arrows scale the lateral flows.

Conceptually, $K_e \ll 1$ means that the basin is much narrower than R_i, that is, a dynamically narrow basin where rotation effects will not be immediately obvious, like a typical fjord. On the other hand, $K_e > 1$ means that the basin is wider than R_i, in such a way that rotation effects should be evident. A situation in which $K_e \ll 1$ and $E_k \ll 1$ (dynamically narrow and deep basins) does not necessarily mean that *Coriolis* acceleration is irrelevant because it would be much more prominent than friction.

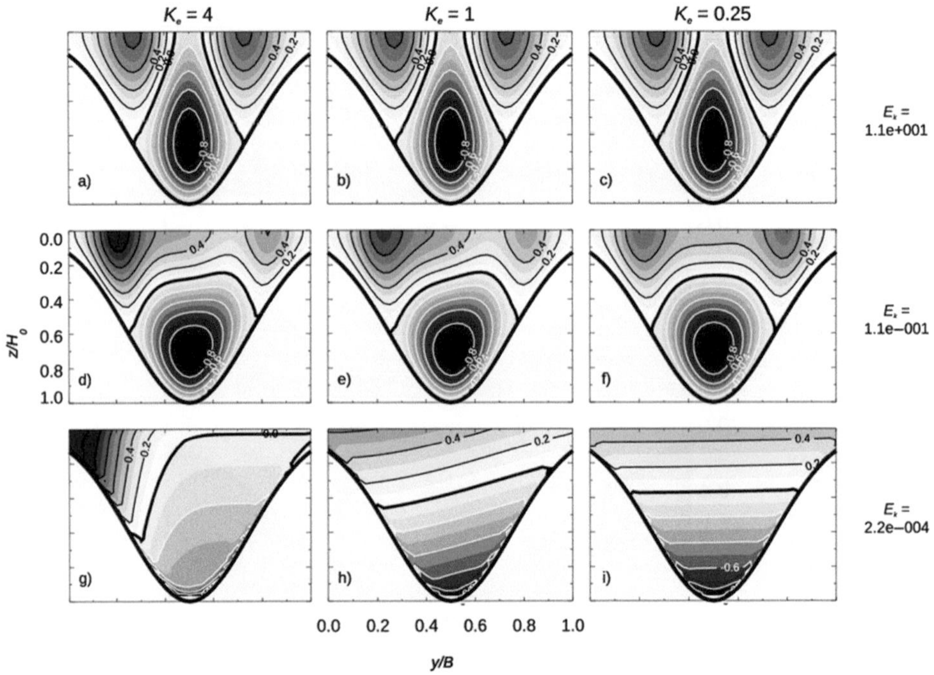

Figure 7.13 Density-driven exchange flows for different values of E_k and K_e. View is landward for the northern hemisphere and seaward for the southern hemisphere. Positive values (blue) represent outflow. All flows have been normalized to the maximum inflow.

More possibilities of E_k and K_e combinations are explored in Figure 7.13 by trying, for fixed E_k, different values of ς (or R_i^{-1}) in equation 7.44. The upper row of panels (a, b, c) in the figure displays the solution for a high E_k (~10), that is, when friction overwhelms rotation. For this linear formulation, the width of the system is irrelevant as the exchange flow is laterally sheared with inflow extending the entire depth of the channel. Rotation effects are unapparent, and the bathymetry shapes the exchange flow. In Section 8.3, we compare these linear results with nonlinear effects from advective flux of momentum.

For intermediate E_k (~0.1, panels d, e, f in Figure 7.13), width indeed becomes relevant. For the wide basin results ($K_e = 4$), the outflow is stronger on its right branch (northern hemisphere) and the inflow is shifted to its right. For the narrow basin ($K_e = 1/4$), *Coriolis* is imperceptible as two cores of near-surface outflow develop on either side of the cross-section. For the frictionless case ($E_k \sim 2 \times 10^{-4}$), the dynamics are geostrophic and the width of the basin determines whether the exchange flow is laterally sheared (Figure 7.13g) or vertically sheared (Figure 7.13i). In these frictionless cases, the slope of the interface of no motion follows Margules' equation. All these solutions apply also to the southern hemisphere, but the view would be seaward instead of landward.

Many semienclosed basins where the mean exchange flow is dominated by density gradients can be placed in the context of the E_k and K_e parameter space. Most estuaries with relatively shallow depths (low E_k) will be found somewhere around the top two rows of Figure 7.13. Dynamically deep semienclosed basins like gulfs, rias, and fjords will occupy the lower row when their forcing is dominated by baroclinicity.

Equations 7.41 and 7.43 may seem menacing but should be tamable with a little patience. To conquer the possible threat from such equations, you will find a Matlab script that draws equation 7.41 for given H, E_k, K_e, and N, with $R = 0$. Trying different values of the parameters will provide a better understanding of the solution and associated processes.

7.5 Take-Home Message

The dynamical framework that yields gravitational circulation consists of pressure gradient balanced by friction in the along-basin direction, and geostrophy in the cross-basin direction. This framework can include lateral bathymetry changes to explain laterally and vertically sheared exchange flows. Adding *Coriolis* and friction to the dynamical framework allows casting solutions in terms of dynamic depth (*Ek*) and width (*Ke*). This framework is linear and also describes whether water exchange is laterally or vertically sheared. Further complications to the framework are described in Chapter 8.

Appendix 7.1

Let us try to fit a mean velocity profile u_0 to a theoretical profile u like the one portrayed by equation 7.13:

$$u(z) = \underbrace{\frac{3}{2}\frac{\boldsymbol{R_w}}{H}\left[1 - \frac{z^2}{H^2}\right]}_{\text{River-induced}} + \underbrace{\frac{gH^3}{48\rho_0 \boldsymbol{A_z}}\frac{\boldsymbol{\partial\rho}}{\boldsymbol{\partial x}}\left[9\left(1 - \frac{z^2}{H^2}\right) - 8\left(1 + \frac{z^3}{H^3}\right)\right]}_{\text{Density-induced}}$$
$$+ \underbrace{\frac{1}{4}\frac{\boldsymbol{\tau_s} H}{\rho_0 \boldsymbol{A_z}}\left[4\left(1 + \frac{z}{H}\right) - 3\left(1 - \frac{z^2}{H^2}\right)\right]}_{\text{Wind-induced}},$$

which can be rewritten in terms of the free parameters, in bold, separately from the constants and depth-dependencies, as

$$u = \underbrace{AZ_1}_{\text{River-induced}} + \underbrace{BZ_2}_{\text{Density-induced}} + \underbrace{CZ_3}_{\text{Wind-induced}}, \tag{A7.1}$$

where

$$Z_1 = \frac{3}{2H}\left[1 - \frac{z^2}{H^2}\right], \quad Z_2 = \frac{gH^3}{48\rho_0}\left[9\left(1 - \frac{z^2}{H^2}\right) - 8\left(1 + \frac{z^3}{H^3}\right)\right], \text{ and}$$
$$Z_3 = \frac{1}{4}\frac{H}{\rho_0}\left[4\left(1 + \frac{z}{H}\right) - 3\left(1 - \frac{z^2}{H^2}\right)\right]. \tag{A7.2}$$

Similarly, $A = \boldsymbol{R_w}$, $B = \boldsymbol{\partial\rho}/\boldsymbol{\partial x}/\boldsymbol{A_z}$, and $C = \boldsymbol{\tau_s}/\boldsymbol{A_z}$. These substitutions are done conveniently to simplify the algebra when we try to minimize the squared error ε^2:

$$\varepsilon^2 = \sum (u_0 - u)^2 = \sum \left(u_0^2 - 2u_0 u + u^2\right), \tag{A7.3}$$

where all sums are over the total number of observations. Substituting u from equation A7.1 into equation A7.3, the squared errors may be written as

$$\varepsilon^2 = \sum \left[u_0^2 - 2u_0AZ_1 - 2u_0BZ_2 - 2u_0CZ_3 + A^2{Z_1}^2 + 2ABZ_1Z_2 \right.$$
$$\left. + 2ACZ_1Z_3 + 2BCZ_2Z_3 + B^2{Z_2}^2 + C^2{Z_3}^2\right], \quad \text{(A7.4)}$$

minimizing the error ε^2 with respect to each of the free parameters A, B, C (not too tough):

$$\frac{\partial \varepsilon^2}{\partial A} = \sum[-2u_0Z_1 + 2A{Z_1}^2 + 2BZ_1Z_2 + 2ACZ_1Z_3] = 0$$

$$\frac{\partial \varepsilon^2}{\partial B} = \sum[-2u_0Z_2 + 2AZ_1Z_2 + 2CZ_2Z_3 + +2B{Z_2}^2] = 0$$

$$\frac{\partial \varepsilon^2}{\partial C} = \sum[-2u_0Z_3 + 2AZ_1Z_3 + 2BZ_2Z_3 + +2C{Z_3}^2] = 0.$$

This system of equations can be written in matrix form, after eliminating all 2s and isolating terms involving the observed profile u_0:

$$\begin{bmatrix} \sum(u_0Z_1) \\ \sum(u_0Z_2) \\ \sum(u_0Z_3) \end{bmatrix} = \begin{bmatrix} \sum {Z_1}^2 & \sum Z_1Z_2 & \sum Z_1Z_3 \\ \sum Z_1Z_2 & \sum {Z_2}^2 & \sum Z_2Z_3 \\ \sum Z_1Z_3 & \sum Z_2Z_3 & \sum {Z_3}^2 \end{bmatrix} \begin{bmatrix} A \\ B \\ C \end{bmatrix}. \quad \text{(A7.5)}$$

Solving this system for A, B, C (see Appendix 4.1) provides the optimal parameters to match equation 7.13 to an observed mean velocity profile u_0. A Matlab script is provided to carry out this calculation. The script has data of a mean velocity profile. It also calculates the optimal parameters and plots the theoretical solution (equation 7.13) together with the observed profile. In addition, the script compares an observed mean salinity profile to the theoretical profile given by equation 7.19.

Additional Sources

Burchard, H., R.D. Hetland, E. Schulz, and H.M. Schuttelaars (2011) Drivers of residual estuarine circulation in tidally energetic estuaries: Straight and irrotational channels with parabolic cross section. *J. Phys. Ocean.* 41: 548–570.

Huntley, H.S., and P. Ryan (2018) Wind effects on flow patterns and net fluxes in density-driven high-latitude channel flow. *J. Geophys. Res.* 123: 305–323.

Kasai, A., A.E. Hill, T. Fujiwara, and J.H. Simpson (2000). Effect of the Earth's rotation on the circulation in regions of freshwater influence. *J. Geophys. Res.* 105(16): 961–969.

Valle-Levinson, A. (2008) Density-driven exchange flow in terms of the Kelvin and Ekman numbers. *J. Geophys. Res.* 113: C04001.

Whitney, M.M., and Y. Jia (2020) Solutions for subtidal flow in channels and estuaries under different integral constraints. *J. Geophys. Res.: Oceans* 125: e2020JC016076.

8

Interactions among Tides, Density Gradients, and Wind

The previous three chapters introduce attributes of residual or tidally averaged flows driven by tides, winds, and density gradients. Flows are treated separately, or, as in the case of density-driven flows that also included wind stress as surface boundary condition, independent of other forcings. In that approach, the contribution from each forcing is added linearly to each other to produce a flow driven by different forcings. This can sometimes be acceptable. However, the reality is that, for example, tides can modify the density gradients at different time scales and these alterations to the density gradient ultimately drive the residual flow. Moreover, winds can modify the density gradients that drive residual flows. Winds can even affect tidal behavior in semienclosed basins when they are strong enough.

This chapter considers interactions between different forcings. It first describes the interaction between tidal currents and density gradients at intratidal (within one tidal cycle) time scales. One outcome of this interaction is the phenomenon known as *tidal straining*. The chapter continues with the treatment of intratidal variations of density that can also result from the interaction of density fields with tides and bathymetry. Subsequently, the chapter presents a description of the interaction between tides and density gradients at subtidal time scales, that is, at periods greater than one tidal cycle. The chapter then describes how advective accelerations from tidal currents can interact with density gradients to modify residual flows. It follows with a description of the competition between tidal stresses and density gradients in driving residual flows. It then deals with the competition between density gradients and wind stresses, to later add tidal forcing. The chapter then includes the influence of river discharge on estuarine circulation. The last two subsections present salt (or solute) budgets and their linkage to hydrodynamics and approaches to study saltwater intrusion.

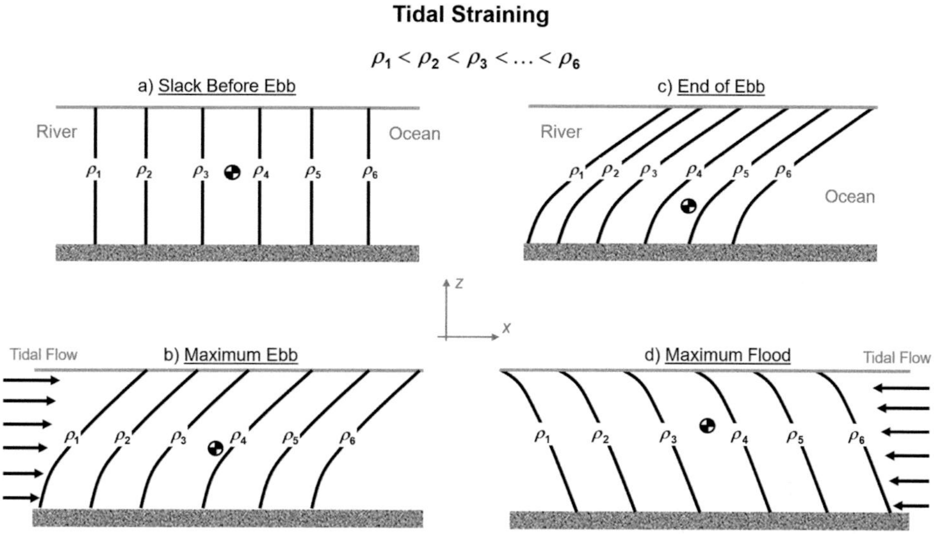

Figure 8.1 Illustration of the concept of tidal straining through isopycnal excursions in the horizontal and vertical directions. The monochromic disc indicates the approximate depth of the water column's center of mass at the profile where it is drawn.

8.1 Intratidal Interactions between Tides and Density Gradients

Interactions between tidal currents and the density field in a semienclosed basin may effect variations in water column stratification. These stratification variations will develop over relatively uniform along-basin bathymetry and morphology, but also over rapidly changing along-basin bathymetry and morphology. Changes in morphology are represented by sills, hollows, or coastline contractions. This section first addresses interactions over gently changing or uniform morphology, and then over rapidly changing morphology.

8.1.1 Interactions over Uniform Morphology

In the simplest case, let us consider a vertically mixed water column but influenced by a horizontal density gradient $\partial\rho/\partial x$ (Figure 8.1a). In this situation, which may occur at the end of flood (or slack before ebb), the horizontal density gradient is depth independent. As the tidal phase progresses toward ebb currents, we expect the tidal current to be vertically sheared because of bottom friction, as shown in Figure 8.1b. The vertical gradients in tidal flow should transport more water and further seaward at the surface than at depth and *strain* the density field in such a way that relatively lighter water overlies heavier water during this phase (Figure 8.1b). As ebb currents continue to flow, the horizontal density gradient is further strained. Maximum straining is reached at the end of ebb (or slack before flood, Figure 8.1c), which is the expected time for greatest vertical stratification. During flood currents, the vertically sheared tidal flow would be expected to transport more water and further into the basin at the surface than at depth (Figure 8.1d). This situation would result in a statically unstable water column that would overturn within minutes, that is, mix vertically, and return to the situation at the end of flood (Figure 8.1a). This sequence of straining of the density field by the vertically sheared tidal currents is known as *tidal straining*.

Tidal straining will produce increased stratification during ebb periods and maximum stratification at the end of ebb. In turn, destratification will ensue during flood, reaching minimum stratification by the end of flood. This variability in stratification will be observed in numerous estuaries, but also there will be several examples where stratification changes differently in a tidal cycle. Even within a basin, you could find tidal straining at some locations but a different behavior at other sites. For instance, maximum stratification could develop during flood because of the intrusion of salty water underneath a homogeneous, relatively fresher water column that was transported seaward during ebb. This situation of increased stratification during flood is also referred to as *reverse tidal straining* because maximum stratification timing is the opposite to that of tidal straining. Reverse straining is frequently found in salt-wedge estuaries or at the upstream limit of saltwater intrusion in other estuaries.

From a quantitative perspective, we can describe changes in water density ρ overt time as the interaction between tidal currents $u_t(z)$ with the horizontal density gradient $\partial\rho/\partial x$, which can be written as a simplified version of conservation of mass (e.g., equation 2.9):

$$\frac{\partial\rho}{\partial t} = -u_t\frac{\partial\rho}{\partial x}. \tag{8.1}$$

This is obviously an approximation as it neglects lateral gradients and flows, as well as vertical advection of a vertical gradient. Equation 8.1 also disregards diffusive effects related to vertical mixing. Still, this approximation expresses the essence of tidal straining if we look at the relative changes of density with depth, that is, differentiating equation 8.1 with respect to z, results in

$$\frac{\partial}{\partial t}\left[\frac{\partial\rho}{\partial z}\right] = -\frac{\partial u_t}{\partial z}\frac{\partial\rho}{\partial x}. \tag{8.2}$$

The left-hand side of equation 8.2 describes changes of stratification $\partial\rho/\partial z$ in time. These stratification variations are given by the interaction between the vertically sheared tidal currents $\partial u_t/\partial z$ and the depth-independent horizontal density gradient, that is, tidal straining. As we can infer from equations 8.1 and 8.2, the temporal

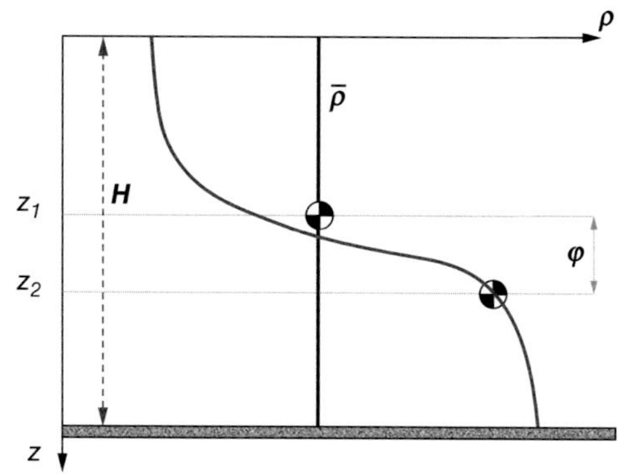

Figure 8.2 Center of mass, C_m, for a mixed and a stratified water column. For a mixed water column, C_m is at $z_1 = H/2$. As the water column becomes stratified, its C_m migrates downward to z_2. The C_m position is denoted by a bicolored disk. The Greek letter φ indicates the difference in potential energy (per unit volume) between a mixed water column with density $\bar{\rho}$ and a stratified water column. φ is also known as the *potential energy anomaly*.

changes of stratification in a semienclosed basin will be caused by tidal straining plus other processes (not included in equations 8.1 and 8.2) that are explained in this section.

An additional, relatively simple, and general approach to examine temporal changes in stratification is through exploration of the *center of mass*, C_m, in a water column. The C_m (in meters) is a metric that condenses conditions of vertical stratification into one number. The C_m of a mixed water column of depth H is in the middle, that is, at $H/2$ (z_1 in Figure 8.2). In a stratified water column, the C_m migrates downward, toward the bottom, to depths greater than $H/2$ (z_2 in Figure 8.2). Values of C_m at any given time can be calculated from a series of M number of water density measurements ρ_i at corresponding depths z_i for a given water column (e.g., from a CTD profile). Values of C_m at any given time and at any horizontal position x, y in a semienclosed basin, are computed as

$$C_m = \frac{\sum_{i=1}^{M} \rho_i \times (H + z_i)}{\sum_{i=1}^{M} \rho_i}. \tag{8.3}$$

Keep in mind that z in equation 8.3 is negative, representing distance from the surface. Two examples illustrate variations of $C_m$within a tidal cycle (Figure 8.3a). One example illustrates profiles of water-density anomalies ($\sigma_t = \rho_t - 1{,}000$) throughout one tidal cycle in a Gulf of Mexico estuary, St. Andrew Bay in Florida.

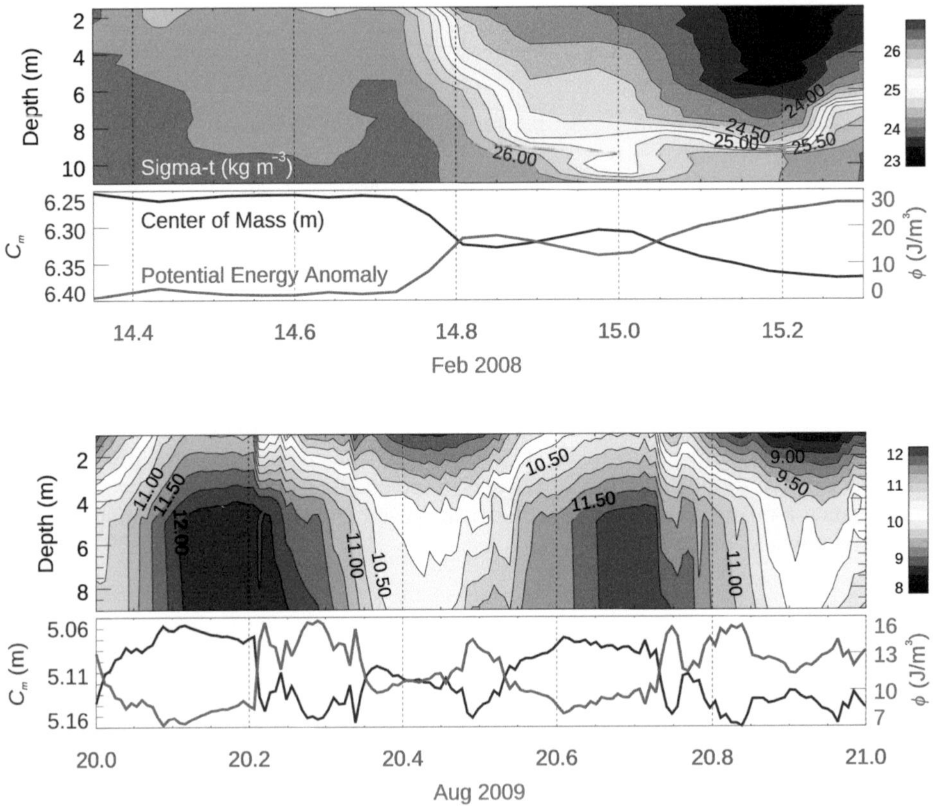

Figure 8.3 Density anomaly profiles throughout one day, together with values of C_m (as in equation 8.3) and potential energy anomaly (as in equation 8.6) at two sites. Both sites exemplify tidal straining behavior in which the greatest stratification appears toward the end of ebb. St. Andrew Bay is dominated by diurnal tides, while James River displays semidiurnal variability. Note that C_m values increase downward (as the C_m goes deeper).

During the flood period, salinity is higher and less stratified, while the water column becomes more stratified during ebb. Values of C_m change a few centimeters throughout the cycle, being deeper in the water column during the stratified ebb periods.

The second example reveals the tidal progression of density anomaly profiles in the James River estuary, a tributary to Chesapeake Bay (Figure 8.3b). These measurements also describe tidal straining variations in stratification, now throughout two tidal cycles. Highest densities are associated with weakest stratification and upward excursions of C_m. Values of C_m could be normalized by the water column H to describe stratification changes in a nondimensional parameter in which higher values would denote higher stratification. These parameters derived from the C_m of a water column can describe variations in stratification but cannot explain the causes for such variations.

Another quantitative approach to study changes in stratification considers the potential energy Π (= mass × gravity × depth = mgH) of the water column (in Joules). In essence, the approach examines deviations of potential energy per unit volume (Π/volume in Joules/m^3) with respect to a mixed water column. Thus, the potential energy per unit volume is $\rho g H$, but because ρ changes with depth z, the comprehensive expression for potential energy per unit volume (J/m^3) is

$$\Pi_V = \frac{g}{H}\int_{-H}^{0} \rho\, z\, dz. \tag{8.4}$$

Consequently, the potential energy per unit volume for a homogeneous water column $\overline{\Pi_V}$ of mean density $\bar{\rho}$ may be written as

$$\overline{\Pi_V} = \frac{g}{H}\int_{-H}^{0} \bar{\rho}\, z\, dz. \tag{8.5}$$

The difference in Π_V between mixed and stratified water columns (J/m^3) is

$$\overline{\Pi_V} - \Pi_V = \frac{g}{H}\int_{-H}^{0} (\bar{\rho} - \rho) z\, dz = \varphi. \tag{8.6}$$

This is known as the *potential energy anomaly* and is another measure of water column stratification, also illustrated in Figure 8.3. This quantity describes the energy (per unit volume in J/m^3) required to mix a water column, that is, the conversion from kinetic to potential energy necessary to destroy stratification. When the water column is mixed, φ equals zero (no energy needed to mix the water column). The greater the value of φ, the stronger the vertical stratification. Yet another way to envision φ is that it represents the energy required to move the C_m of a water column to its half-depth point, that is, move C_m from its stratified position to $H/2$ (see also Figure 8.2).

A formalism to study changes of stratification φ in time, caused mostly by interactions between tidal currents and the density field, considers equation 8.6 and its changes over time, that is,

$$\frac{\partial \varphi}{\partial t} = \frac{g}{H}\int_{-H}^{0} \left(\frac{\partial \bar{\rho}}{\partial t} - \frac{\partial \rho}{\partial t}\right) z\, dz. \tag{8.7}$$

Units in equation 8.7 are Watts per cubic meter (energy per unit time and volume, W/m^3). Note that to understand the evolution of stratification $\partial \varphi/\partial t$ (in W/m^3), we need expressions for $\partial \rho/\partial t$ and for $\partial \bar{\rho}/\partial t$. For the former, we can use the equation for density or transport equation for density (analogous to conservation of salt, equation 2.20):

$$\frac{\partial \rho}{\partial t} = -u\frac{\partial \rho}{\partial x} - v\frac{\partial \rho}{\partial y} - w\frac{\partial \rho}{\partial z} + \frac{\partial}{\partial x}\left(K_h\frac{\partial \rho}{\partial x}\right) + \frac{\partial}{\partial y}\left(K_h\frac{\partial \rho}{\partial y}\right) + \frac{\partial}{\partial z}\left(K_\rho\frac{\partial \rho}{\partial z}\right) + S_b, \tag{8.8}$$

where $S_b(x, y, z, t)$ represents sources and sinks of buoyancy that drive density changes such as river (even rain) input or evaporation. Equation 8.8 describes changes in water density over time as being caused by advective processes (first three terms on the right hand), horizontal diffusive processes (following two terms), vertical diffusive processes (next term), and buoyancy processes, which can be positive or negative.

The next requirement to assess changes in stratification according to equation 8.7 is to provide a statement for depth-averaged density changes $\partial\bar{\rho}/\partial t$. For such statement, the density and velocities in equation 8.8 are assumed to have a depth mean plus a vertical deviation from their depth means, that is, $\rho = \bar{\rho} + \rho'$, $u = \bar{u} + u'$, and $v = \bar{v} + v'$. The expression for $\partial\bar{\rho}/\partial t$ is then obtained by vertical integration of equation 8.8 from the bottom ($z = -H$) to the surface ($z = 0$), following equation 8.7, and using the mean plus deviations substitution, which yields

$$\frac{\partial(\bar{\rho}H)}{\partial t} + \frac{\partial(\bar{u}\bar{\rho}H)}{\partial x} + \frac{\partial(\bar{v}\bar{\rho}H)}{\partial y} = -\frac{\partial\left(\overline{u'\rho'}H\right)}{\partial x} - \frac{\partial\left(\overline{v'\rho'}H\right)}{\partial y} + \left(K_\rho\frac{\partial \rho}{\partial z}\right)_{z=0} - \left(K_\rho\frac{\partial \rho}{\partial z}\right)_{z=-H} + \overline{S_b}. \tag{8.9}$$

Remember, we are trying to explain changes in stratification over time, considering potential energy anomalies. Equation 8.9 together with 8.8 are inserted into equation 8.7 to explain the processes responsible for stratification changes anywhere in the coastal ocean:

$$\frac{\partial \varphi}{\partial t} = \frac{g}{H}\int_{-H}^{0}\left\{\begin{array}{l}\underbrace{\bar{u}\frac{\partial \rho'}{\partial x} + \bar{v}\frac{\partial \rho'}{\partial y}}_{I} + \underbrace{u'\frac{\partial \bar{\rho}}{\partial x} + v'\frac{\partial \bar{\rho}}{\partial y}}_{II} + \underbrace{u'\frac{\partial \rho'}{\partial x} + v'\frac{\partial \rho'}{\partial y}}_{III} \underbrace{- \frac{1}{H}\frac{\partial(\overline{u'\rho'}H)}{\partial x} - \frac{1}{H}\frac{\partial(\overline{v'\rho'}H)}{\partial y}}_{IV} \\ \underbrace{w\frac{\partial \rho}{\partial z}}_{V} \underbrace{- \frac{\partial}{\partial z}\left(K_\rho\frac{\partial \rho}{\partial z}\right)}_{VI} \underbrace{- \frac{\partial}{\partial x}\left(K_h\frac{\partial \rho}{\partial x}\right) - \frac{\partial}{\partial y}\left(K_h\frac{\partial \rho}{\partial y}\right)}_{VII} \underbrace{+ \frac{1}{H}\left(K_\rho\frac{\partial \rho}{\partial z}\right)_{z=0} - \frac{1}{H}\left(K_\rho\frac{\partial \rho}{\partial z}\right)_{z=-H}}_{VIII}\end{array}\right\} z\,dz \tag{8.10}$$

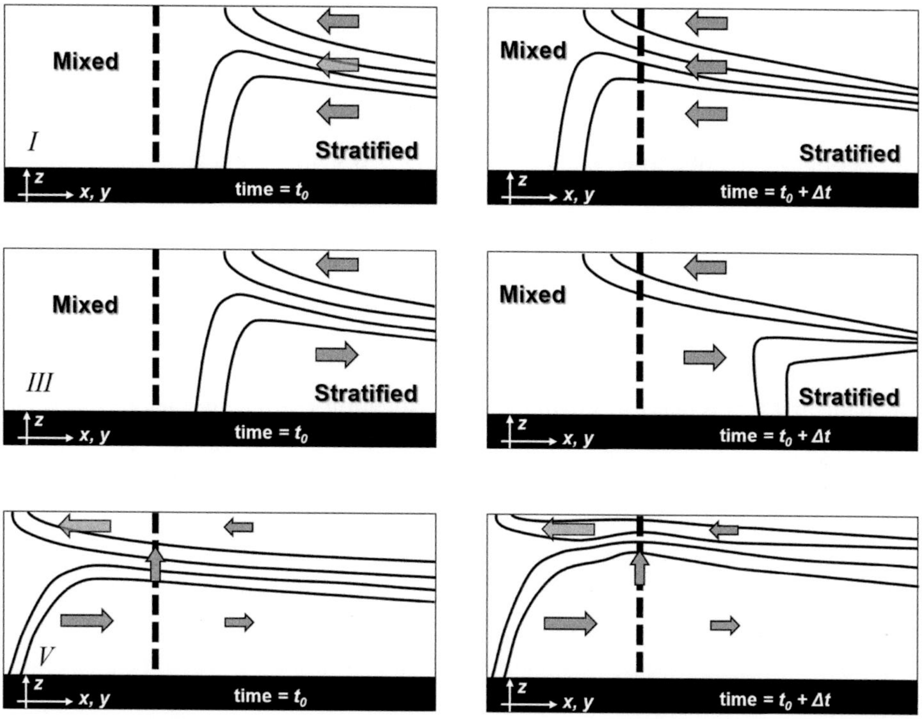

Figure 8.4 Schematic of mechanisms *I*, *III*, and *V* from equation 8.10. Mechanism *II* is illustrated in Figure 8.1. The diagrams illustrate changes in stratification from time t_0 (on the left frames) to time $t_0 + \Delta t$ (on the right frames) at the location of the thick vertical dashed line.

At first sight, this expression appears unwieldy. However, it is quite accessible. It identifies eight mechanisms (*I* through *VIII*) responsible for changes in stratification over time (W/m^3) for any water column of depth *H*. Except for *V*, *VI*, and *VIII*, these mechanisms have two horizontal components in the *x* and *y* directions. Mechanisms *I* through *V* are illustrated schematically in Figure 8.4.

Mechanism *I* represents *advection of a stratified water column* by depth-mean flows, as in a tidal current advecting a stratified plume, without distorting it, to cause changes in stratification (top two panels of Figure 8.4). Mechanism *II* embodies the interaction between vertically sheared flow and a depth-independent horizontal density gradient, that is, *tidal straining* (Figure 8.1). Mechanism *III* characterizes the interaction between a depth-dependent horizontal density gradient and a vertically sheared current, similar to tidal straining but depth-dependent. Mechanism *IV* involves the horizontal gradients of the vertical covariance between flow shears and density stratification. It is the depth-averaged version of Mechanism *III* and will contribute to changes in stratification in instantaneously heterogeneous (in *x* or *y*) density and flow fields.

Mechanism *V* describes upward or downward advection of stratification related to upwelling and downwelling. Mechanism *VI* describes vertical mixing, while mechanism *VII* denotes horizontal mixing. Mechanism *VIII* characterizes the fluxes of buoyancy at the surface and bottom. In areas where tidal amplitude is non-negligible relative to H, that is, where $\eta/H > 0.1$, the integrals in equation 8.10 should go to η, the variations in water level, instead of going to 0. In this case, two other terms should be considered for changes in stratification:

$$\underbrace{-\frac{\varphi}{H}\frac{\partial H}{\partial t}}_{IX} \underbrace{-\frac{g}{H}\rho'\eta\frac{\partial \eta}{\partial t}}_{X}. \tag{8.11}$$

Mechanism *IX* is related to changes in stratification caused by changes in water column depth and mechanism *X* is linked to changes in surface elevation. In equations 8.10 and 8.11, terms with negative signs represent decrease in stratification when they themselves are positive. Also in equations 8.10 and 8.11, positive values of $\partial\varphi/\partial t$ indicate increased water column stratification.

These are the equations that permit a full description of changes in stratification. Because of their apparent complexity, the contribution of each mechanism (*I* through *X*) can be assessed with numerical model results. It would be an exciting challenge to try to decipher each mechanism observationally. Keep in mind that in many cases, studies concentrate on *tidal straining* (mechanism *II*) as the main driver of stratification changes. Equations 8.10 and 8.11 indicate the need to consider several other drivers. The influence of other mechanisms becomes much more evident when maximum stratification occurs at a different phase from the end of ebb. And even in that circumstance, there is a need to determine the role of vertical mixing (mechanism *VI*).

It is evident that intratidal variations in stratification can provide completely different vertical distributions of solutes and suspended materials depending on the phase of the tidal cycle. Any sampling effort that disregards this type of variability in semienclosed basins will necessarily be biased or incomplete.

8.1.2 Interactions among Tidal Currents, Density Field, and Morphologic Changes

We concentrate now on along-basin morphologic changes related to sills and hollows, and to coastline constrictions. Before we describe interactions of a stratified flow with bathymetry, we will begin by using an analogy that describes a simpler situation. Consider first a homogeneous flow moving over an obstacle where bed friction is negligible as in Figure 8.5, animated in Video 8.1. In general,

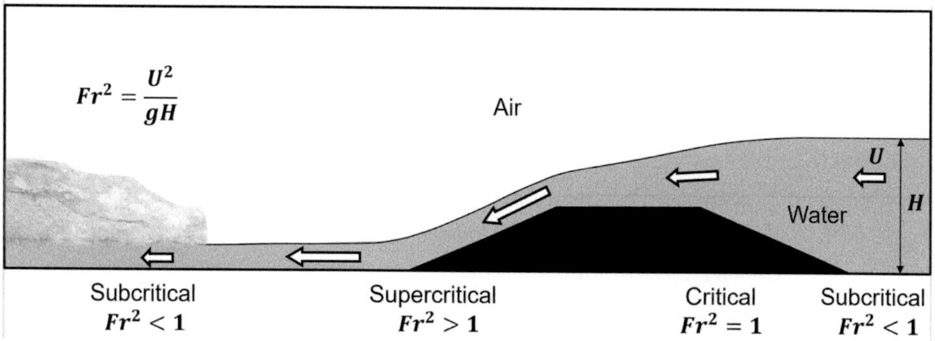

Figure 8.5 Schematic (horizontal and vertical plane) of the two-dimensional flow over an obstacle changing from subcritical to critical and supercritical and back to subcritical. The transition from supercritical to subcritical flow occurs through a hydraulic jump with enhanced turbulence and mixing of momentum.

this situation is analogous to wind moving over a mountain, or through a mountain gap, or even through the entrance to a subway station.

The interaction between the flow and the obstacle, or reduction in cross-sectional area through which the fluid moves, can be described from the perspective of conservation of mechanical energy. In other words, kinetic plus potential energy must conserve in the domain. The flow moves toward the obstacle at speed U throughout the depth H (Figure 8.5). As the flow moves past the obstacle, it accelerates and its kinetic energy increases because it is proportional to the squared of its velocity. With the kinetic energy increasing, the potential energy has to decrease to conserve energy. Potential energy declines through vertical migration of the water column's center of mass toward the bottom while H decreases (Figure 8.5). The point where the flow speed equals the speed of the long gravity wave ($\sqrt{gH}$) is where the flow is said to be *critical* and where the water surface begins to drop. As the flow accelerates even further while going downslope, the kinetic energy grows even further with a corresponding drop in the water elevation (Figure 8.5). Throughout this region, $U > \sqrt{gH}$ and the flow becomes *supercritical* (Figure 8.5). Transitions from *subcritical* ($U < \sqrt{gH}$) to supercritical and back to subcritical flow are described with the Froude number ($Fr^2 = U^2/gH$ or $Fr = U/\sqrt{gH}$). This nondimensional number arises from a Bernoulli-type momentum balance that compares advective fluxes to pressure gradients. It is also a comparison of kinetic to potential energies.

Downstream of the obstacle the flow decelerates and the water elevation suddenly jumps upward at the transition from supercritical ($Fr > 1$) to subcritical ($Fr < 1$) flow. This sudden increase in depth is called a *hydraulic jump*. The jump in surface elevation launches a wave with susceptibility to propagate at phase speed $\sqrt{gH}$ against the flow. In a hydraulic jump the wave continually breaks and is unable to propagate against the flow as $U = \sqrt{gH}$. If $U > \sqrt{gH}$ the jump will be washed downstream, whereas if $U < \sqrt{gH}$ the jump-related wave will propagate upstream (against the flow). A relevant phenomenon associated with the hydraulic jump is the turbulence and mixing of momentum and mass that ensues from the breaking wave (Figure 8.5 and Video 8.1).

The same processes, that is, transitions from subcritical to supercritical and back to subcritical flows, can be identified in river flows moving over a rock (Video 8.2). Likewise, hydraulic transitions are expected as flows move through coastline constrictions and through depressions or hollows in the bathymetry. Video 8.2 demonstrates fixed hydraulic jumps downstream of obstacles (rocks). Examples in Figure 8.5 and Videos 8.1 and 8.2 are for homogeneous, unidirectional flows. However, in semienclosed basins influenced by density gradients, the flow will most likely be oscillatory and stratified.

A two-layered fluid influenced by an oscillatory flow and bathymetry displays similar behavior to that described above. The interface between air and water for the case of a homogeneous layer has a density difference of order 10^3 kg/m^3 (Figure 8.5). In comparison, the interface between two layers in the water column typically has a density contrast of order <10 kg/m^3 (maximum contrast in the interior of the water column can be around 25–30 kg/m^3). But the interface behavior in the latter is analogous to that described in Figure 8.5. The behavior of an interface (e.g., pycnocline) in a stratified fluid interacting with an obstacle can be described with the composite Froude number (Fr_c), which is the sum of the Froude number in the upper and lower layers (Figure 8.6).

$$Fr_c^2 = Fr_1^2 + Fr_2^2 = \frac{U_1^2}{gH_1} + \frac{U_2^2}{gH_2}, \tag{8.12}$$

where U_1, U_2 are the flow speeds representative of the upper and lower layers, respectively, while H_1, H_2 are their corresponding depths. An example of this situation is illustrated in Figure 8.6. Data were collected along Paso Galvarino in the Chilean Inland Sea, within the Puyuhuapi Fjord. The sampling site features a sill and contraction morphology. Over 2 km along the fjord, the bathymetry changes landward from a depth of 80 m to ~10 m at the sill and down to 40 m landward of the sill. Over the same landward tract, the coastline contracts from 2 km to 100 m (close to the shallowest point) and then expands again to 2 km. In the example of Figure 8.6, the pycnocline is relatively close to the surface, centered at a depth of ~5 m. During flood flows, the pycnocline bends down at the dramatic cross-section reduction while the flow accelerates (Figure 8.8a and c). At the same time, acoustic backscatter increases with flow speed and values of Fr_c^2 go through one hydraulic transition from subcritical to supercritical and back to subcritical (Figure 8.8b). Backscatter increases markedly at the zone landward of the transition

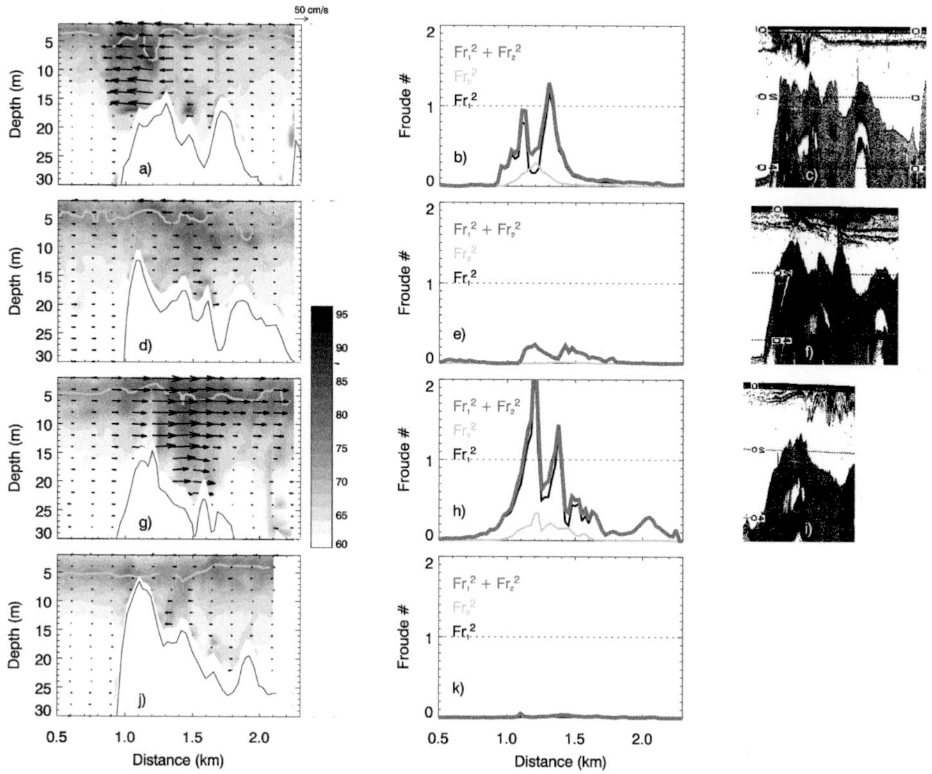

Figure 8.6 Observations of hydraulic transitions throughout an along-basin transect at a sill/contraction combination in a Chilean fjord, Paso Galvarino. The rows display fields at (from top to bottom) maximum flood, end of flood, maximum ebb, and end of ebb. The left column plots velocity fields (black arrows), acoustic backscatter (shades) and bathymetry (continuous line). The middle column illustrates layer and composite Froude number. The right column displays echosounder traces corresponding to the two frames to their left. No trace was available for the bottom transect crossing, at the end of ebb.

from supercritical to subcritical, that is, landward of the hydraulic jump where increased turbulence and mixing are expected. Enhancement of vertical exchange of properties landward of the hydraulic jump is also shown on an echosounder trace (Figure 8.6c) by the pycnocline becoming increasingly blurry. In this example, the upper layer dominates values of Fr_c^2 as it is the "active" layer responding to morphologic changes. In other basins, the lower layer is the active layer or both layers may contribute similarly to values of Fr_c^2.

By the end of flood (Figure 8.6d–f), the flow begins ebbing near the bottom while it is still flooding near the surface, as discussed in Section 3.4.5. Flows are subcritical everywhere and the pycnocline remains relatively flat. During maximum ebb (Figure 8.6g–i), the tidal flow is stronger than during maximum flood and there are two hydraulic transitions from supercritical to subcritical. The pycnocline excursions associated with these transitions are evident on the echosounder trace, which displays two oscillations of order 100 m in length. Seward of the second oscillation, the trace exhibits a diffuse pycnocline, which indicates enhanced mixing. By the end of ebb (Figure 8.6j and k), there are no hydraulic transitions anymore as the flows remain subcritical. The flow begins to flood near the bottom while it still ebbs at the surface (e.g., Section 3.4.5).

Changes in stratification along a domain with an obstacle or a constriction can be studied with equation 8.10 to understand the processes responsible for such variations. The best approach to address this topic presently is with numerical simulations. Existing technology is still unavailable to determine every term of equation 8.10 reliably with measurements. Moreover, any attempt to determine all terms in the field would be extremely demanding in human and instrumental resources.

A major consequence of the interactions between stratified fluids and oscillatory (tidal) flows through morphologic changes (sills, hollows, or constrictions) is to cause differences between flood and ebb processes. The flood–ebb asymmetry in processes becomes apparent in long-term or residual fields. These morphologic changes represent "hot spots" for mixing processes that can promote primary productivity locally. They also promote processes similar to those around headlands, as described in Section 5.5. Essentially, flow accelerations and decelerations associated with morphologic changes trigger processes that enhance and dissipate property gradients, with corresponding biogeochemical responses.

8.2 Interactions between Tides and Density Gradients at Subtidal Scales

As mentioned in the previous paragraph, intratidal processes are usually asymmetric between flood and ebb phases. These asymmetries translate into non-tidal (or residual, or subtidal) fields. Expected responses in semienclosed basins consist of increased vertical mixing during periods of enhanced tidal currents at spring tides in semidiurnal regimes and at tropic tides in diurnal regimes. This variability is commonly known as fortnightly variability. As seen in Section 3.3, we should distinguish between *synodic* and *declinational* fortnight. Synodic fortnight is also associated with the spring–neap tidal cycle with a period of 14.77 d and dominates in semidiurnal tides. Declinational fortnight, or simply fortnight, is associated with the tropical–equatorial cycle of the Moon's declination with a period of 13.66 d and appears in tides with diurnal dominance.

During the strongest tidal currents of the fortnight, density gradients ($\partial\rho/\partial x$ and $\partial\rho/\partial z$) are expected to be weakest, driving the feeblest exchange flows (Figure 8.7a). Similarly, the greatest exchange flows and maximum stratification

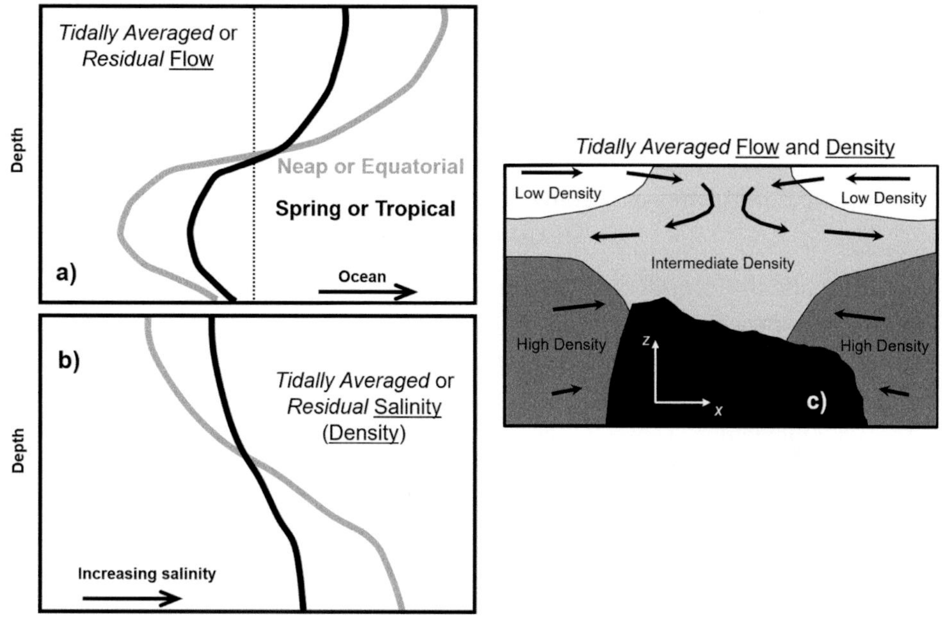

Figure 8.7 Expected tidally averaged profiles of exchange flow (a) and salinity (b) for maximum and minimum tidal ranges in one fortnight. The vertical axes are depth and the horizontal axes are residual flow or salinity (density). (c) Residual flows and density (salinity) field expected from increased vertical mixing over a local shoal or obstacle interfering a stratified flow. Note the resulting three-layered structure. Half of this panel (left or right) would be relevant to a situation with enhanced vertical mixing at the head of a stratified tributary.

are expected during periods of the weakest tidal currents (Figure 8.7a). The influence of fortnightly tidal current on exchange flow and stratification can be assessed, at first approximation, with equations 7.13 for exchange flow and 7.19 for salinity profile. Both equations indicate that larger gradients and lower eddy viscosities, during weakest tidal currents, result in increased exchange flows and stratification in the mean salinity (density) profile (Figure 8.7b). Profiles schematized in Figure 8.7 are for conditions when density gradients dominate the tidally averaged flows and tides modulate the influence of density gradients.

At subtidal scales, the interaction of a stratified, oscillatory flow with an obstacle will favor vertical mixing over the obstacle (Figure 8.7c). Vertically mixed water of intermediate density, a mixture of low- and high-density waters, will be surrounded by stratified water columns. The intermediate density water will then tend to intrude between the buoyant and the heavy layers, resulting in a three-layered flow. The residual flow over such obstacle will thus exhibit convergence, and accumulation of suspended and dissolved material, at surface and bottom of the obstacle. Divergence will ensue at the depth of the pycnocline at either side of the obstacle. This long-term mechanism of near-surface convergent flows at a shoal should favor feeding and nursing grounds for species of commercial and ecological importance.

8.3 Influence of Tidally Related Advective Accelerations on Density-Driven Flows

Tidal flows produce residual circulations in instances of marked gradients in tidal currents like those generated by lateral variations in bathymetry (Section 5.2 and Figure 5.2, and Section 8.4). But gradients in tidal currents can also modify density fields through the mechanisms depicted in Section 8.1 and through advective fluxes of momentum. These fluxes do not necessarily produce residual currents that overcome density-driven flows but have a dynamical influence. When this happens, the tidally averaged momentum balance is nonlinear, different from equation 7.2, as it must include a residual advective flux of momentum:

$$\left\langle u_T \frac{\partial u_T}{\partial x} \right\rangle + \left\langle v_T \frac{\partial u_T}{\partial y} \right\rangle + \left\langle w_T \frac{\partial u_T}{\partial z} \right\rangle = -g \frac{\partial \langle \eta_T \rangle}{\partial x} - \left\langle \int_{-H}^{z} \frac{g}{\rho_0} \frac{\partial \rho_T}{\partial x} dz \right\rangle + \left\langle A_z \frac{\partial^2 u_T}{\partial z^2} \right\rangle. \tag{8.13}$$

In equation 8.13, brackets $\langle\rangle$ indicate tidal averages and the subindex $_T$ denotes tidal properties. Making the left-hand side equal to zero turns this equation into 7.2. Qualitatively, the influence of the added advective momentum flux is to reinforce the vertically sheared, density-driven exchange flow with outflow at the surface and inflow underneath. However, this is expected under certain bathymetric cross-sections, which have been explored only with numerical models.

Findings over a given cross-section can be cast in the parameter space of the Ekman number (E_k, i.e., dynamic depth – Section 7.4 and equation 7.38) and a dynamic width. A dynamic width is formed from the ratio of advective to *Coriolis* accelerations, adding the latter to the momentum balance in equation 8.13. Such ratio yields the Rossby number ($R_o = U/fB$, where U is a typical magnitude of the exchange flow, f is the *Coriolis* parameter and B is basin's width). Comparing Figure 8.8 to Figure 7.12, the residual advective momentum flux seems to be most influential in relatively narrow ($R_o > 1$) and shallow ($E_k > 0.1$) basins (upper-right panel in Figure 8.8). That is the region in the parameter space where the linear exchange flow changes from laterally sheared (Figure 7.12) to vertically sheared (Figure 8.8). In the rest of the parameter space, the contribution of advective fluxes to exchange flows seems to be minor as there is essentially no qualitative difference between linear and nonlinear dynamics. Thus, outside of dynamically shallow and narrow basins, density-driven exchange flows are dominated by

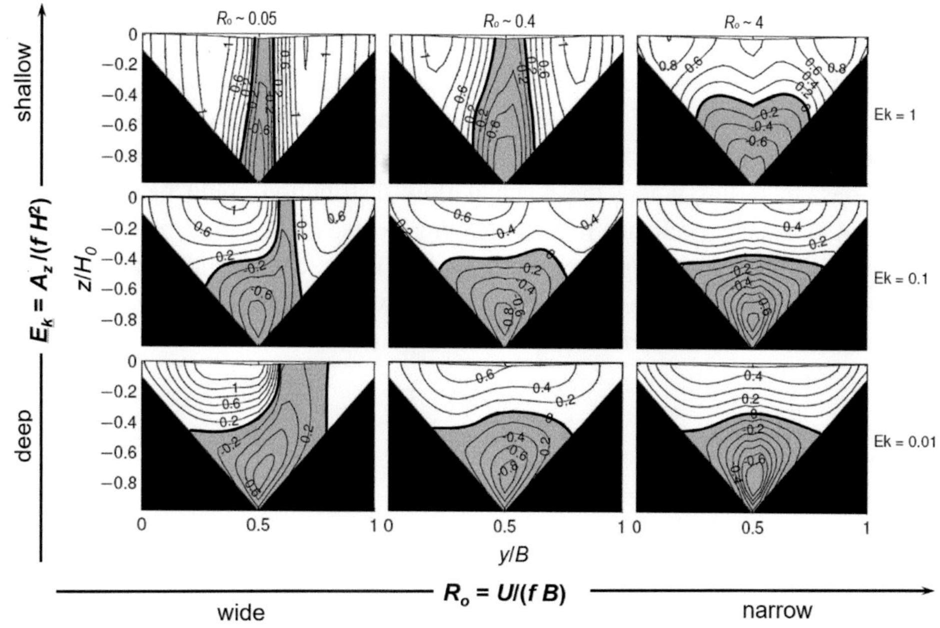

Figure 8.8 Density-driven exchange flows for different values of dynamic depth E_k and dynamic width R_o. Results were derived from numerical simulations with fully implemented advective accelerations. View is landward for the northern hemisphere and seaward for the southern hemisphere. Positive values (white contours) represent outflow. All flows have been normalized to the maximum inflow. Results provided by Peng Cheng.

Coriolis and frictional accelerations balancing the pressure gradient. Additional exploration is needed on the effects of advective accelerations on density-driven flows for other bathymetric configurations.

An additional consequence of tidal flows interacting with density gradients arises from the stress divergence terms (last term in equation 8.13). This term has a component that represents the covariance between the tidal variations of the eddy viscosity A_z and the vertical shears in tidal currents. This term has been called ESCO (eddy–shear covariance) and is denoted by $\langle A_{zT} \partial u_T / \partial z \rangle$, where the subindex T indicates tidal variations. The ESCO mechanism has been found to reinforce the gravitational circulation, and in some instances overwhelm it. Further information on this mechanism is being developed and goes beyond the scope of this text.

8.4 Competition between Tidal Stress and Density-Gradient in Driving Residual Flows

Previous sections describe the interaction between tidal and density-gradient forcings as the tide modifies or modulates the residual circulation driven mainly by density gradient. There are basins, however, where tidal stress actually competes with baroclinicity to ultimately determine the force that drives exchange flows. From previous sections, we have seen that tidally induced residual circulation in "short" basins consists of inflow in channel and outflow over shoals, which coincides with expected density-driven flows under dominantly frictional conditions. In basins where the residual flow distribution could be ascribed to tides and to density gradients, it is challenging to assess what force actually drives it. One way of addressing such a challenge is by comparing tidal stress to baroclinicity. Tidal stresses arise from advective accelerations, which may be scaled in terms of tidal current amplitudes U_0, as

$$\left\langle u_T \frac{\partial u_T}{\partial x} \right\rangle \sim \frac{U_0^2}{L_T}, \tag{8.14}$$

where L_T is the tidal length scale, that is, the *tidal excursion*. Values of tidal excursion L_T may be obtained by integrating an oscillatory motion (the tide) over one half period ($T/2$), for example:

$$L_T = \int_0^{T/2} U_0 \sin\left(\frac{2\pi}{T}t\right) dt = \frac{U_0 T}{\pi}. \tag{8.15}$$

For a semidiurnal ($T = 12.42$ h) tidal current amplitude U_0 of 1 m/s, L_T equals approximately 14 km.

Returning to the comparison of tidal stress to baroclinicity, the baroclinic pressure term may be scaled in terms of the reduced gravity, g', as

$$\left\langle \int_{-H}^{z} \frac{g}{\rho_0} \frac{\partial \rho_T}{\partial x} dz \right\rangle \sim \frac{g}{\rho_0} \frac{\Delta\rho}{L_\rho} H = \frac{g' H}{L_\rho}, \tag{8.16}$$

where L_ρ is the length over which the density gradient is determined, e.g., saltwater intrusion length. Comparison of the tidal stress scaling over baroclinicity scaling, equation 8.14 over 8.16, yields

$$Fr_t = \frac{L_\rho}{L_T} \frac{U_0^2}{g' H} = \gamma \frac{U_0^2}{g' H}, \tag{8.17}$$

which we refer to as the *densimetric Tidal Froude number* or simply the *Tidal Froude* number. The factor γ is the ratio between density-gradient length scale and tidal length scale. Typically, this value should be between orders 1 and 10 in semienclosed basins.

An example of the competition between tidally driven and density-driven flows comes from data collected in a tropical estuary in northeastern Brazil, the Mossoro. The northeastern Brazilian region is characterized by strong seasonality in rain

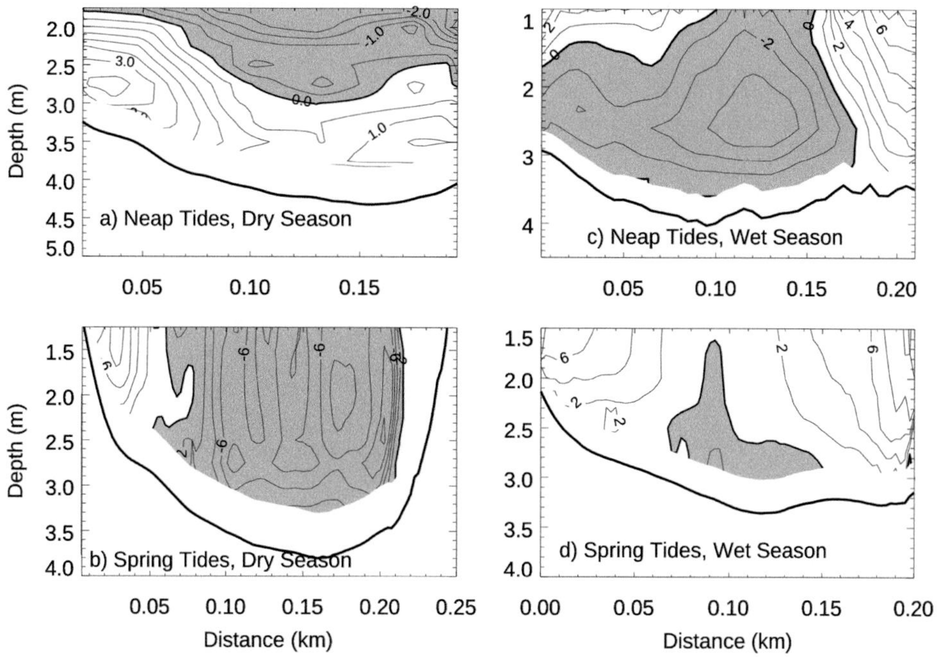

Figure 8.9 Fortnightly switching in forcing for exchange flows. Data were collected with a towed ADCP during diurnal cycles in Mossoro estuary, northeast Brazil. Observations were collected in the dry season (a, b) and the wet season (c, d). Views are into the estuary, with gray contours denoting residual inflows. The black line shows the bathymetry profile.

dominated by arid, evaporative conditions. During the eight-month-long dry season, the basin behaves as an *inverse*, *hypersaline*, and *hyperpycnal* (see Chapter 11) estuary with maximum salinities reaching 70 g/kg. During the four-month-long wet season, the estuary displays normal estuarine behavior with salinity decreasing monotonically landward. Tidal amplitudes are 1.5 m in spring tides and 0.75 m in neaps over average depths between 2 and 3 m. Both tidal forcing and baroclinicity are therefore expected to drive residual flows.

In the dry season, the spatial structure of exchange flows changes dramatically from neap to spring tides. In neap tides, inverse estuarine conditions favor vertically sheared exchange flow with inflow in a surface layer and outflow underneath (Figure 8.9a). In spring tides, however, the exchange flow is laterally sheared with inflow in the channel and outflow over shoals (Figure 8.9b). This flow structure is consistent with that expected for tidal residual flows in "short" basins, but opposite to that expected for hyperpycnal density-driven flow under increased frictional influence. Therefore, it is likely that the residual flow is driven by density gradients in neap tides and by tidal forcing in spring tides.

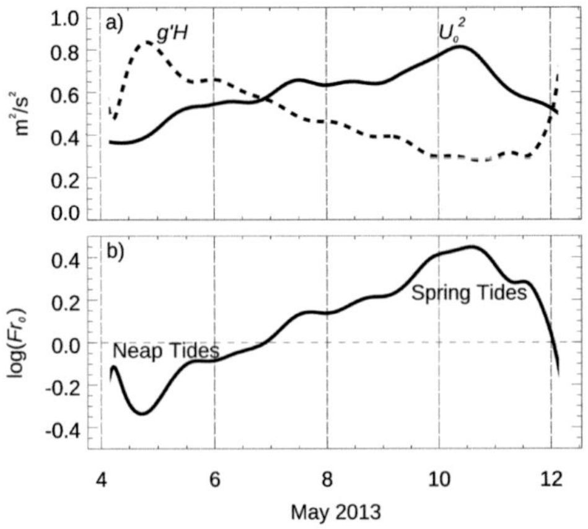

Figure 8.10 Densimetric tidal Froude number over a seven-day period to illustrate the fortnightly switching of forcing. Numerator and denominator in equation 8.17 (a) display expected fortnightly variability. The same can be said about values of log (Fr_t).

The possibility of fortnightly switching in forcing is bolstered by what happens in the wet season (Figure 8.9c and d). In spring tides (Figure 8.9d), the exchange flow is laterally sheared although dominated by a river discharge pulse that developed before the measurements. This is consistent with both tidally driven and density-driven flows. During neap tides (Figure 8.9c), the exchange flow is more vigorous than during spring tides as it is more likely driven by density gradients than in spring tides.

Fortnightly switching in forcing was confirmed with week-long time series in the wet season that allowed tracing of U_0^2, $g'H$, and Fr_t in the basin (Figure 8.10). During the wet season, $g'H$ are maxima in neap tides and decrease toward spring tides. Tidal current amplitudes follow the opposite trend. Values of γ in equation 8.17 are around 1 and Fr_t can be plotted in a logarithmic scale (base 10). This transformation contracts ratios when the numerator in equation 8.16 is too big and expands the ratios when the denominator is lower than 1 (negative logarithm). Thus, values of log (Fr_t) are negative in neaps, indicating baroclinicity dominance, and positive in spring tides, denoting tidal stress dominance.

Equation 8.17 surely applies in other systems to diagnose the main driver of residual flows. This has been shown in Chelem, a costal lagoon in the Yucatan Peninsula; in the Gironde, a macrotidal estuary in western France; and in the Guadiana, an estuary that separates southern Portugal from southern Spain. In contrast to a case in which $Fr_t < 0.1$ where only density gradients dominate the exchange flow, as in Section 8.2, when $Fr_t > {\sim}2$, the stronger exchange flows will appear in spring tides and not in neap tides.

8.5 Competition between Density Gradients and Wind Forcing

The next interaction or competition to explore is between density gradients and wind forcing. As mentioned in Chapter 6, wind forcing can mix the water column, drive water-level slopes, and move water. This section concentrates on how wind can enhance, hinder, or even dominate over the density-driven exchange flows.

The competition between wind stress τ_s and baroclinicity may be scaled as

$$\frac{\tau_s}{\rho_0 H} \; vs. \; \frac{g}{\rho_0}\frac{\Delta\rho}{L_\rho}H, \tag{8.18}$$

resulting in the Wedderburn number *We*:

$$We = \frac{\tau_s L_\rho}{g\,\Delta\rho H^2}. \tag{8.19}$$

When values of *We* are between 0.2 and 2 (exact values remain to be determined), wind will modify density-driven flows but not overwhelm them. Within this range, in semienclosed basins influenced by fresh water, seaward winds should enhance density-driven exchange flows and strengthen density vertical stratification. Landward winds should act in the opposite direction, reducing stratification and density-driven flows. Modification of vertical stratification by wind stress advection has been coined *wind straining*, which can act both to enhance or to deter stratification.

The response of density-driven flows to different wind forcings is illustrated with numerical results that neglect tidal forcing (Figure 8.11). When *We* is small, around 0.2, the exchange flow is enhanced as the wind blows seaward and delayed as the wind acts landward (against the surface outflow). The exchange flow pattern is altered only in few detailed features. However, when *We* is large, that is, >5, the exchange flow becomes essentially wind-driven: downwind over shoals and upwind in the channel (see also Chapter 6). Remember that the discussion of these concepts has disregarded the influence of tidal forcing, which is considered in the following section.

8.6 Addition of Tidal Forcing to Wind and Density-Gradient Interactions

Tidal forcing, much the same as wind forcing, modifies density gradients through moving gradients, mostly horizontally (advection), and via smoothing gradients, mainly vertically (vertical mixing). As seen in previous sections, the modifications can have periodicities from minutes to fortnight to month and year, depending on the forcing. Besides, the modifications are complicated by bathymetry and by nonlinear mechanisms in the momentum balance, such as advective fluxes and

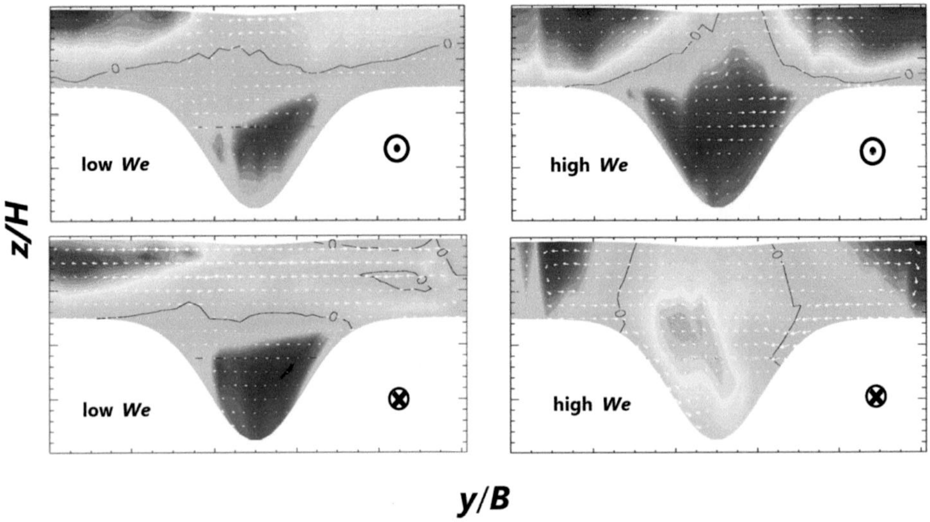

Figure 8.11 Numerical results of density-driven flows as modified by wind stress. Simulations included *Coriolis* accelerations. View is into the basin (Northern Hemisphere) with normalized distance across the basin (y/B) and normalized depth (z/H). Red contours indicate outflow (toward the viewer) with the interface drawn as black contour. Density-driven flows are slightly altered under low *We* scenarios. High *We* scenarios illustrate predominantly wind-driven flows. Wind blowing toward the viewer is indicated by a circle with a dot in the middle. Winds away from the viewer are represented by a circled "X." White arrows illustrate transverse flows. No scale is shown for either of the flow components because results portray only qualitative distributions.
Results provided by Rosario Sanay

stress divergence. This chapter for the most part disregards those nonlinear mechanisms, which make interactions among tides, wind, density gradients, and bathymetry different from a linear superposition of each process. Understanding such interactions is an active area of research that requires increased efforts. The complexity of such interactions is illustrated with numerical model results of residual or tidally averaged fields that include the three main forcing agents: tides, density gradients, and wind.

A reference exchange flow field with no wind forcing in the middle of an estuary (Chesapeake Bay in this case) serves to epitomize the complex interactions (Figure 8.12 from numerical simulations). In the reference state (months-long average), outflow occurs at the surface and occupies most of the shoals, inflow is restricted to depths greater than the mean depth. Addition of wind forcing from four different quadrants, representing *We* around 1, causes distinct responses. Winds toward the mouth, in general and as mentioned on the previous section, enhance density-driven exchange flows. However, the angle of the wind relative to

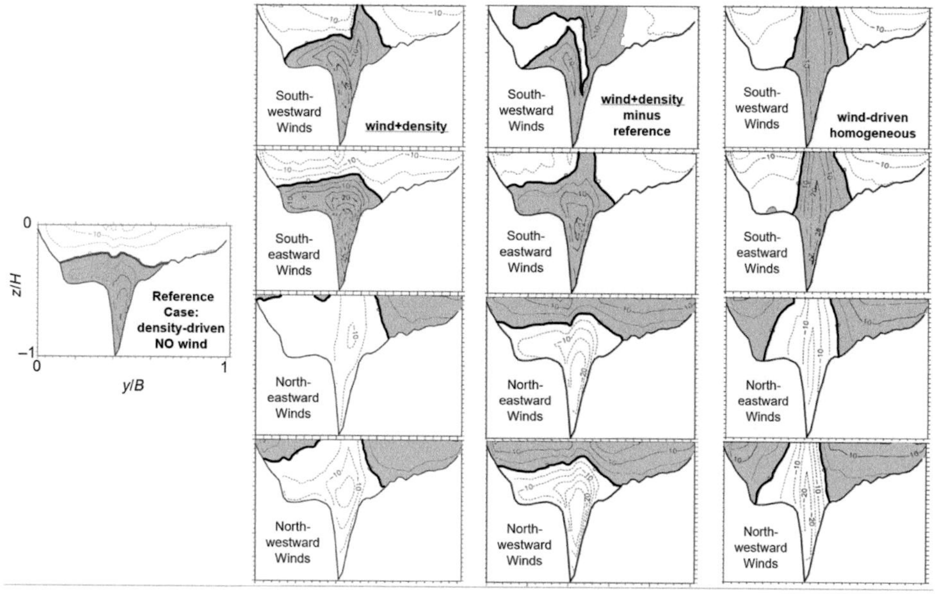

Figure 8.12 Wind modifications to density-driven flows in the presence of tidal forcing. Numerical results at a cross-section in the middle of the Chesapeake Bay (looking into the estuary). Gray-shaded contours indicate inflows. Contour labels are in cm/s. In general, winds with a southward component are seaward. The reference case has with no wind forcing.
Results provided by Xinyu Guo

the axis of the estuary determines whether upwind flows reach the surface. On the other hand, winds toward the estuary's head cause downwind flows over preferentially one shoal and upwind in the channel, reversing the density-driven circulation (Figure 8.12).

The complexity of the density gradient–wind tides interactions is portrayed by subtracting the reference density-driven exchange flow field (without wind forcing) from each one of the fields modified by wind forcing (Figure 8.12). If interactions were linear, this subtraction would yield the same flow fields as those obtained for wind forcing over a constant-density estuary (Figure 8.12). Wind-driven flow fields in a homogeneous basin show downwind flows over shoals and upwind flows in channel (as in Chapter 6). These wind-driven flow fields under homogeneous water, as expected, are different from the wind modifications to the density-driven flows. Generalities on wind modifications to density-driven flows in the presence of tidal forcing and bathymetry remain a fundamental question in the study of semienclosed basins.

Observations of wind modifications to the gravitational circulation under variable tidal forcing illustrate the influence of up-estuary winds (Figure 8.13), that is, winds blowing against the near-surface outflow. Portions of these observations

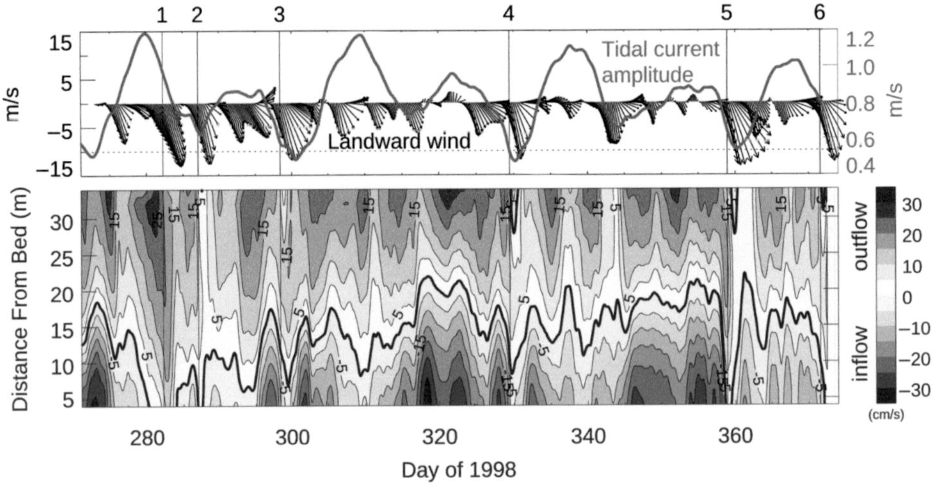

Figure 8.13 Subtidal (low-pass filtered) velocity profiles modified by wind forcing in a Chilean estuary, Meninea Strait. Upper panel shows wind velocity. Its sign is the direction toward which it blows, with negative values indicating landward winds. The orange line describes variations in tidal current amplitudes with a well-defined synodic fortnightly cycle. In the lower panel, the subtidal flows (cm/s) are positive toward the ocean and negative toward the head of the estuary. The thick contour is the interface between outflow and inflow.

have been used in Figures 6.2a, 7.1, and 7.3b. Subtidal (low-pass filtered) velocity profiles described exchange flows over a sill in a semienclosed basin of the Chilean Inland Sea for almost 100 d. The basin is roughly oriented in the N–S direction, with the entrance at its northern end. During the measurements period, there were seven fortnightly cycles and six wind pulses that exceeded 10 m/s (Figure 8.13). Positive subtidal flows near the surface indicated outflow.

The overall trend of subtidal flows in the record was for outflow near the surface and inflow underneath. The fortnightly tidal cycle had insignificant effects in the exchange flows throughout the record. However, the typical exchange flow was disrupted by the six up-estuary wind pulses that weakened the outflow, but also the inflow. The last three wind pulses (4, 5, and 6; Figure 8.13) even drove inflow at the surface, a water level set-up toward the head (not shown) and a seaward near-bottom flow that counteracted the gravitational inflow. Weakening of exchange flows by up-estuary winds was signaled by the downward excursion of the interface between inflows and outflows (thick contour).

8.7 Influence of River Discharge on Density-Driven Flows

Another fundamental question in the study of hydrodynamics in semienclosed basins is what happens to density gradients and exchange flows with increasing river discharge. This is perhaps the most basic question for which a general response is elusive. We know that density gradients generate gravitational circulation. But the answer to what happens to that circulation by increasing river discharge is "it depends"! It actually depends on the basin length, width, and depth, on the temporal scales involved, and on the size of the freshwater pulse.

In the simplest description of the response of density-driven flow to river discharge, we can examine equation 7.13 and Figure 7.2 (also equation 7.37 in 7.9 and Figure 7.10 with R, instead of R_w). From the simplified dynamics, we can glean that river discharge causes domination of volume outflow and decrease volume inflow. Salinity values thus tend to decrease everywhere in the basin. However, in a real scenario, it is possible that horizontal salinity gradients are enhanced either inside or outside the semienclosed basin. A response could be that after the initial freshwater pulse, most likely after relaxation of the freshwater pulse, the gravitational circulation is enhanced by the strengthened horizontal salinity gradients. Another response in an actual situation, such as in a deep fjord, is that increased discharge is restricted to a surface layer, detached from the bottom, that drives bolstered inflow underneath the seaward layer. Circumstances that cause enhanced or hindered gravitational circulation with augmented river discharge still lack universality. This has been a focus topic of investigations for decades and continues to require full attention. One way of addressing this topic is through the study of salt fluxes in particular, and solute fluxes in general.

8.8 Salt and Solute Budgets: Linkage to Hydrodynamics

Salt fluxes are carried out through different mechanisms that are discussed in this section. Understanding the response of a semienclosed basin to river discharge variations may hinge in recognizing the dominant mechanisms that drive the transport of salt. Through a relatively straightforward calculation we can learn about the mechanisms responsible for transport of salt and solutes.

To study salt fluxes in a semienclosed basin, we can represent values of salinity and along-basin (or axial) velocity component at any point (and at a given time) on a cross-section (e.g., Figure 8.14) as the sum of a cross-sectional average (denoted by an overbar) plus the spatial deviation from that average (indicated by an apostrophe):

$$u = \bar{u} + u' \tag{8.20}$$

$$S = \bar{S} + S'. \tag{8.21}$$

These expressions provide spatial context for the flow and salinity fields at a prescribed time. For instance, the salt transport caused by the *mean flow* $\bar{u}$ is given by

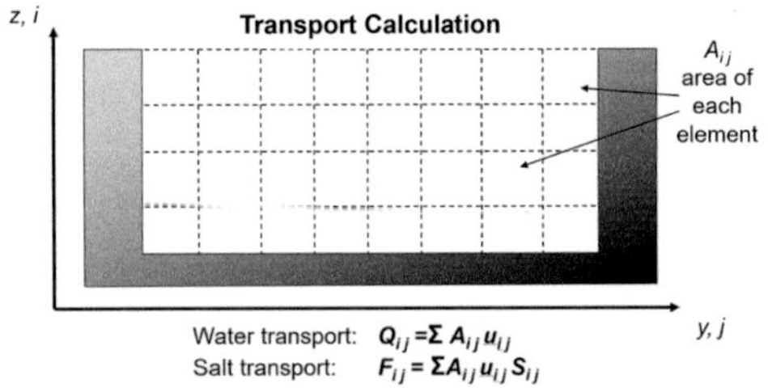

Figure 8.14 Generic cross-section in a semienclosed basin to calculate transport of solutes. The cross-section is partitioned in rows (i) and columns (j). At each area element A_{ij} of the cross-section, there is a value of along-basin velocity u_{ij} and salinity S_{ij}. The transport of mass (water) is given by the sum of all area elements times their corresponding velocities. Salt flux or transport is given by the sum of all water transports times the corresponding salinity in each area element.

$$\textit{mean transport} = \int_A \bar{u} \cdot \bar{S} \cdot dA \approx \sum \bar{u}_{ij} \bar{S}_{ij} A_{ij} = \bar{u}\, \bar{S}\, A, \tag{8.22}$$

where the subindices *i,j* refer to the diagram of Figure 8.14. On the other hand, and following equations 8.20 and 8.21, the salt flux caused by spatial deviations from the mean is referred to as *shear dispersion* flux and expressed by

$$\textit{shear dispersion} = \int_A u' \cdot S' \cdot dA \approx \sum u'_{ij} S'_{ij} A_{ij}. \tag{8.23}$$

Remember that the apostrophe in this equation denotes spatial variations. The *shear dispersion* flux itself may be separated in two components. One component involves vertical variations (or variations in the water column) and is called *vertical shear dispersion*. The second component describes the spatial variations across the estuary and is known as *transverse shear dispersion* (Figure 8.15).

Thus, at any given time, the transport of salt (and of any solute S) through a cross-section of area A in a semienclosed basin is represented by

$$\overline{u\, S A} \approx \sum \overline{u_{ij} S_{ij}} A_{ij} = \sum \bar{u}_{ij} \bar{S}_{ij} A_{ij} + \sum u'_{ij} S'_{ij} A_{ij}. \tag{8.24}$$

Flows u_{ij} include contributions typically from river flow and tidal flow at the time of calculation. Flows could also contain the effects from wind and waves. Each one of these flows, in turn, has intratidal (within a tidal cycle) and subtidal or residual contributions. In order to distinguish subtidal from intratidal contributions, we further decompose flows and salinities in their tidal variation, indicated by a

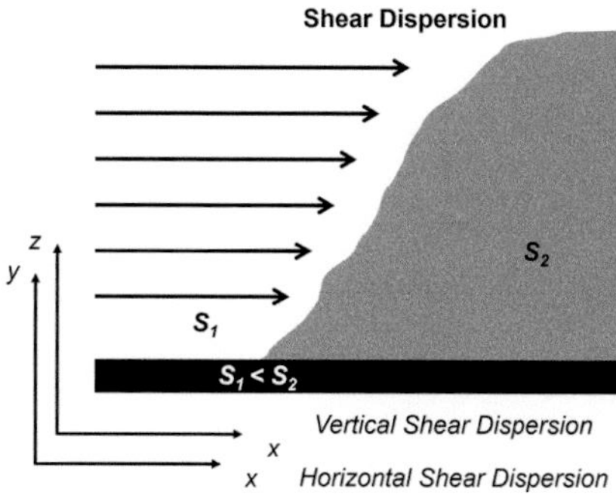

Figure 8.15 Schematic of transport of a solute *S* by shear dispersion. The diagram illustrates vertical and horizontal shear dispersion, according to the set of axes chosen.

tilde ~ and with zero temporal average, plus a subtidal variation, denoted by brackets ⟨ ⟩:

$$\bar{u} = \langle \bar{u} \rangle + \tilde{\bar{u}}; \; u' = \langle u' \rangle + \tilde{u'}$$
$$\bar{S} = \langle \bar{S} \rangle + \tilde{\bar{S}}; \; S' = \langle S' \rangle + \tilde{S'}. \tag{8.25}$$

The net or tidally averaged or subtidal salt transport may then be expressed with the dominant terms (there would be more terms – interactions – that tend to be negligible) as

$$\underbrace{\left\langle \sum \overline{u_{ij} S_{ij}} A_{ij} \right\rangle}_{net} = \underbrace{\left\langle \sum \bar{u}_{ij} \bar{S}_{ij} A_{ij} \right\rangle}_{advection} + \underbrace{\left\langle \sum u'_{ij} S'_{ij} A_{ij} \right\rangle}_{shear\ dispersion} + \underbrace{\left\langle \sum \tilde{\bar{u}}_{ij} \tilde{\bar{S}}_{ij} A_{ij} \right\rangle}_{tidal\ pumping} + \text{interactions}. \tag{8.26}$$

In integral form, equation 8.26 may be represented as

$$\underbrace{\left\langle \int_0^B \int_{-H}^0 u \cdot S \, dz \, dy \right\rangle}_{net} = \left\langle \int_0^B \int_{-H}^0 \left(\underbrace{\bar{u} \bar{S}}_{advection} + \underbrace{u' S'}_{sheared} + \underbrace{\tilde{\bar{u}} \tilde{\bar{S}}}_{pumping} \right) dz \, dy \right\rangle + \text{interactions}. \tag{8.27}$$

In equations 8.26 and 8.27, the transport by *advection* is mainly related to *river discharge* (tidally averaged flow and salinity, sectionally integrated) and/or transport of water from outside the basin (remote transport). If x is positive seaward, this transport will be positive (seaward transport of salt) when it is dominated by river discharge or by seaward transport driven by wind. The *sheared* transport is caused by tidally averaged flows that are vertically sheared or laterally sheared. This

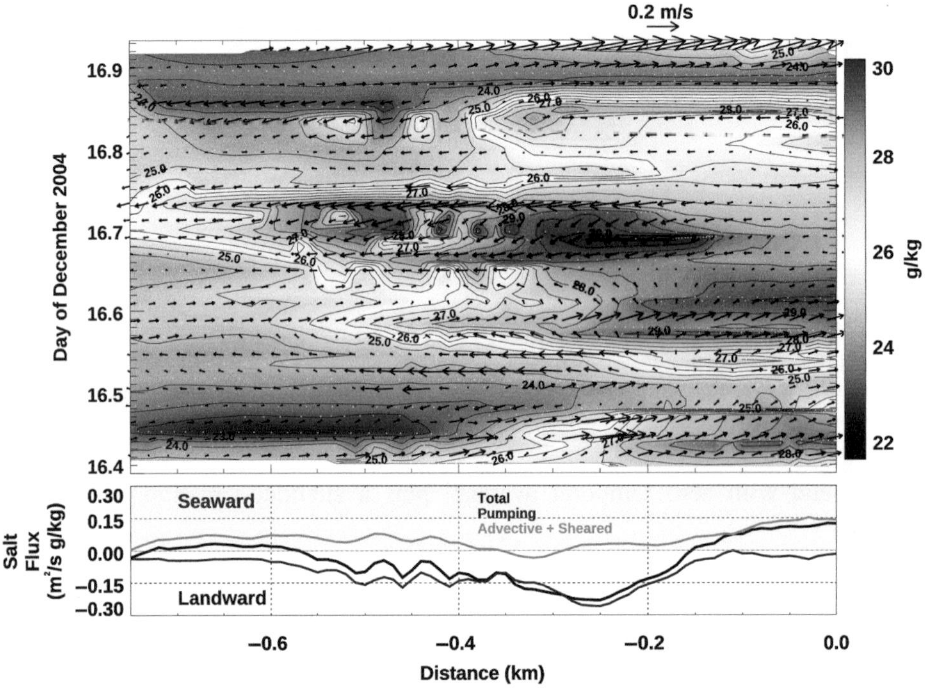

Figure 8.16 Salt flux calculated over a surface layer along a Chilean fjord transect where there are measurements of flow and salinity (color contours). Upper panel shows a Hovmöller diagram of surface-layer velocity and salinity. Lower panel illustrates the distribution of salt fluxes partitions along the transect. Seaward is to the right on both panels. *Total* salt flux equals *tidal pumping* plus *advective* and *sheared dispersion*. Positive fluxes are seaward.

transport will be related to gravitational circulation, but it can also be linked to wind-driven flows or tidal residual flows that are sheared. If the transport is dominated by gravitational circulation and, again, x is positive seaward, this transport will be negative (landward transport of salt).

The *tidal pumping* transport is related to the covariance between tidal flows and tidal fluctuations in salinity. As relatively saltier water typically enters an estuary during flood, this mechanism transports salt into the estuary and will be negative. Tidal pumping depends on the phase relationship between tidal currents and salinity variations. When salinity extremes are reached at times of, or near, slack tidal currents, the tidal variations of salinity and flow will be in quadrature. In this case their covariance will be zero (or near zero) and sheared transport will dominate over tidal pumping. In other words, when salinity and tidal currents are 90° out of phase, the salt flux will be dominated by gravitational circulation. This is the situation in most basins where salt flux calculations have been performed. Tidal pumping should contribute to net salt fluxes when salinity extremes occur at the time of tidal current maxima, as illustrated in Figure 8.16.

In the example of Figure 8.16, the portion of the transect where tidal pumping dominates the salt flux appears at a shoaling of the bathymetry (constriction of the cross-section) where flows accelerate, salinity is advected most effectively, and flow and salinity are in phase (highest salinities occur during strongest tidal floods). Away from the shoaling, the salt flux is as in many other places, where it is governed by the advective and shear dispersion mechanisms. These mechanisms provide information on how saltwater responds to increases in freshwater discharge.

8.9 Saltwater Intrusion

In many basins influenced by freshwater discharge, river pulses are expected to enhance the salt–flux advective mechanism, which drives saltwater seaward. Conceptually, increased river discharge R (in m^3/s) curtails the intrusion of saltwater χ (in m or km) in a semienclosed basin. Values of χ can be defined on the basis of a given isohaline, such that χ_2, for example, represents the position of the 2 g/kg isohaline relative to the mouth of the basin. In basins where χ decreases with increased R, the conceptual (empirical) relationship that implies steady state follows a general power decay:

$$\chi \sim R^{-1/n} \tag{8.28}$$

With $1/n$ representing the rate of decay of salt intrusion with increasing discharge in a given system. This relationship holds in dynamically shallow basins, but more widespread measurements are required. Despite the apparent generality of equation 8.28, there is no universality for the value of n. The approximation given by equation 8.28 can also be represented as

$$\chi = \beta_1 R^{-1/n} + \beta_2. \tag{8.29}$$

Equation 8.29 has three free parameters β_1, β_2, and n, which can be fitted to observations of χ vs. R, following the procedures outlined in the Appendix 7.1. Parameters β_1 and β_2 calibrate values of χ at extreme values of R. Figure 8.17 illustrates arbitrary prescription of the free parameters.

Actual semienclosed basins exhibit markedly different values of n, as illustrated in Figure 8.17. The Suwannee River, on the west coast of Florida, USA, displays values of n between 2 and 3, with relatively short lengths of intrusion (<10 km). Another Florida basin, but on the east coast, the St. Johns estuary, has n values almost 10 times smaller than the Suwannee. Smaller n in this basin results from a relatively small range of river discharge, as the freshwater is derived mostly from

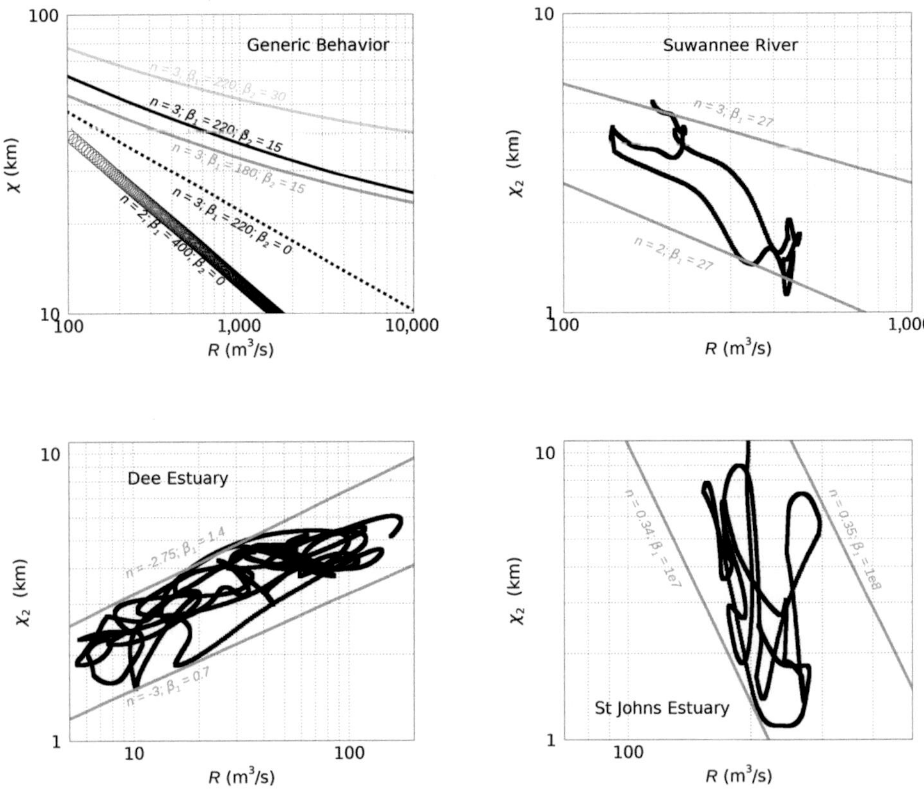

Figure 8.17 Relationships between saltwater intrusion limit χ and river discharge R (equation 8.29). In the panel with a "Generic Behavior," lines illustrate different arbitrary combinations of β_1, β_2, and n. These lines illustrate a decrease in saltwater intrusion with increased river discharge. The three other panels exemplify estuarine systems with dramatically contrasting responses of saltwater intrusion to river discharge. In all three examples, $\beta_2 = 0$.

groundwater discharge. The empirical behavior suggests that lengthy saltwater intrusions should be expected in this system if the discharge decreases from the typically expected values shown, that is, if R becomes <100 m^3/s. Finally, the Dee Estuary, between England and northern Wales, connecting to the Irish Sea, features negative n values as saltwater intrusion increases with river discharge. In basins that feature R versus χ behavior like the Dee, that is, increased intrusion with increased discharge, it is likely that there is a threshold value of R, beyond which saltwater intrusion will begin to decrease as the basin will approach the behavior of a river. These are topics that remain scarcely explored.

Variations of salinity at any given time in a semienclosed basin can be explored with a differential approach anchored on the balance of salt integrated across the basin, and similar to the approach in equations 8.26 and 8.27:

$$\frac{\partial \bar{S}}{\partial t} = -\frac{1}{A}\frac{\partial}{\partial x}\left[\underbrace{\bar{u}\bar{S}A}_{\text{river}} + \underbrace{\overline{u'S'A}}_{\text{sheared}} - \underbrace{K_x \frac{\partial \bar{S}}{\partial x} A}_{\text{tidal diffusive}}\right], \tag{8.30}$$

where the overbar again denotes cross-sectional means and the apostrophe indicates deviations from those means. The "river" term is analogous to the "advection" term in the salt flux expressions. Alternatively, or in addition, to the empirical equation 8.29, we can use equation 8.30 to study the limit of saltwater intrusion as a function of R. The most complete approach, however, would be to use the full form of the salt balance equation (equation 2.20) and solve it numerically over a realistic or semi-realistic domain. Recent efforts have examined salinity variance instead of salinity. Investigations that focus on this topic of salt fluxes, but also on fluxes of any material, will be of fundamental relevance to the management of freshwater resources. Saltwater intrusion is exacerbating at the imminent double threat from sea-level rise and from increased human consumption of fresh water.

8.10 Take-Home Message

This chapter presents topics that represent magnificently fertile ground for research in the hydrodynamics of semienclosed basins. It also combines topics from previous chapters that illustrate the vast and complex variability found in these systems. Any study, in any aspect of a semienclosed coastal body of water, should consider such variability. There is still immense knowledge to gain on interactions between tides and density gradients, both at intratidal (there is more to it than *tidal straining*) and subtidal (there is more than "typical" fortnightly variability) time scales. Also to elucidate are the effects of bathymetry, both along- and across-basin, on the tide–density gradient interaction. Further to complicate understanding, there is the effect of wind that can affect density gradients and tidal forcing. Knowledge of these interactions requires improvement. Finally, increased understanding is essential for the fundamental response of semienclosed basins to river discharge variations and fluxes of dissolved and suspended materials. All processes in this chapter will be altered by expected changes in sea levels. We need to understand how.

Additional Sources

Burchard, H., and R. Hofmeister (2008) A dynamic equation for the potential energy anomaly for analysing mixing and stratification in estuaries and coastal seas. *Estuarine Coastal and Shelf Science* 77: 679–687.

de Boer, G.J., J.D. Pietrzak, and J.C. Winterwerp (2008) Using the potential energy anomaly equation to investigate tidal straining and advection of stratification in a region of freshwater influence. *Ocean Modelling* 22: 1–11.

Geyer, W.R., and P. MacCready (2014) The estuarine circulation. *Ann. Rev. Fluid Mech.* 46(1): 175–197.

Lerczak, J.A., W.R. Geyer, and R.J. Chant (2006) Mechanisms driving the time-dependent salt flux in a partially stratified estuary. *J. Phys. Ocean.* 36(12): 2296–2311.

Lerczak, J.A., W.R. Geyer, and D.K. Ralston (2009) The temporal response of the length of a partially stratified estuary to changes in river flow and tidal amplitude. *J. Phys. Ocean.* 39(4): 915–933.

MacCready, P. (1999) Estuarine adjustment to changes in river flow and tidal mixing. *J. Phys. Ocean.* 29: 708–726.

MacCready, P. (2007) Estuarine adjustment. *J. Phys. Ocean.* 37: 2133–2145.

Monismith, S.G., W. Kimmerer, J.R. Burau, and M.T. Stacey (2002) Structure and flow-induced variability of the subtidal salinity field in northern San Francisco Bay. *J. Phys. Ocean.* 32: 3003–3019.

Scully, M.E., C. Friedrichs, and J. Brubaker (2005) Control of estuarine stratification and mixing by wind-induced straining of the estuarine density field. *Estuaries* 28(3): 321–326.

Scully, M.E., W.R. Geyer, and J.A. Lerczak (2009) The influence of lateral advection on the residual estuarine circulation: A numerical modeling study of the Hudson River Estuary. *J. Phys. Ocean.* 39(1): 107–124.

Simpson, J.H., J. Brown, J. Matthews, and G. Allen (1990) Tidal straining, density currents, and stirring in the control of estuarine stratification. *Estuaries* 13(2): 125–132.

9
Fronts

This chapter describes fronts in, or originating from, semienclosed basins. It begins by proposing a general definition of a front and then presents possible spatial and temporal scales attributable to these features. It proposes arguments that support possible values for vertical velocities at fronts. This chapter discusses different types of fronts arising from the conservation equation for property C, which can be salinity, or density, or any dissolved or suspended material. Such different types of fronts are *plume* and *tidal intrusion* fronts, *mixing* (mostly tidal) fronts, and *shear* fronts. Concepts in the chapter refer to fronts in semienclosed basins and in coastal oceans, although, in some instances, the concepts may be applicable to open ocean regions.

9.1 Definition: Spatial and Temporal Scales

It is challenging to define a *front* in the coastal ocean. It is less of a challenge to see one, like a definition of an *estuary*. In general, we can say that a front is a *rapid* change of a scalar property, or the enhancement of a gradient of such scalar. The term is borrowed from meteorology, but in the ocean the changing scalar may be temperature, salinity, chlorophyll, suspended sediment, dissolved oxygen, or any other dissolved or suspended material, denoted by C. But when we say rapid, what does it really mean? Is it a change of ΔC over a distance Δy? How large should values of $\Delta C/\Delta y$ be to consider them a front?

We can propose that values of $\Delta C/\Delta y$ at coastal ocean fronts should be at least one order of magnitude larger than background values. Such a concept applies to changes in space but could also pertain to changes in time Δt instead of Δy. Consequently, there is no universal definition for a front in coastal waters. However, a quantitative approach to study fronts shall shed light on the mechanisms that enhance gradients. The study of fronts may begin from an equation inspired by conservation of density (equation 8.8). For a scalar property C

its conservation equation consists of advective fluxes of C, diffusive fluxes of C, and sources/sinks of C:

$$\frac{\partial C}{\partial t} = -u\frac{\partial C}{\partial x} - v\frac{\partial C}{\partial y} - w\frac{\partial C}{\partial z} + \underbrace{\frac{\partial}{\partial x}\left(K_h\frac{\partial C}{\partial x}\right)}_{F_x} + \underbrace{\frac{\partial}{\partial y}\left(K_h\frac{\partial C}{\partial y}\right)}_{F_y} + \underbrace{\frac{\partial}{\partial z}\left(K_z\frac{\partial C}{\partial z}\right)}_{F_z} + S_C, \tag{9.1}$$

where diffusive fluxes (throughout a unit volume) have been denoted as F_x, F_y, and F_z in the along-basin x, across-basin y, and vertical z directions, respectively. From equation 9.1, we can study the evolution in time of spatial gradients of C. We concentrate on gradients in the y direction (lateral or transverse direction) in semienclosed basins. In the following, and in other chapters, u is the along-basin flow (in the x direction), v is the lateral flow, and w is the vertical flow. Applying the gradient in y on equation 9.1, we get

$$\frac{\partial}{\partial t}\left\{\frac{\partial C}{\partial y}\right\} = -\underbrace{\left\{u\frac{\partial}{\partial x} + v\frac{\partial}{\partial y} + w\frac{\partial}{\partial z}\right\}\frac{\partial C}{\partial y}}_{translation} + \underbrace{\frac{\partial}{\partial y}\left\{\frac{\partial F_z}{\partial z} + \frac{\partial F_x}{\partial x} + \frac{\partial F_y}{\partial y} + S_C\right\}}_{mixing}$$
$$-\underbrace{\frac{\partial u}{\partial y}\frac{\partial C}{\partial x}}_{shear} - \underbrace{\left\{\frac{\partial v}{\partial y}\frac{\partial C}{\partial y} + \frac{\partial w}{\partial y}\frac{\partial C}{\partial z}\right\}}_{plume}. \tag{9.2}$$

Equation 9.2 describes evolution of gradients in C. The equation says that gradients of a scalar C can increase or decrease over time via four processes: by *translation* of a gradient, by spatial differences in *mixing*, by *lateral shears* of along-basin flows, and by flow *divergence* and spatial gradients in vertical velocities (*plume* fronts). Three out of these four processes generate fronts through enhancement of C gradients. *Translation* moves gradients but in itself is not a mechanism of front formation. Mechanisms that form fronts are referred to as mechanisms for *frontogenesis*. These mechanisms are explored separately in the following sections.

9.2 Plume Fronts and Tidal Intrusion Fronts

Plume fronts (e.g., Figure 9.1) have been the most widely studied and have inspired several review efforts that synthesize their properties. Plumes are typically constrained by an order-meters zone of enhanced horizontal gradients of property C (e.g., density or salinity). Enhanced gradients may be associated with a distinguishable change in color, turbidity or surface roughness, and customarily may be detected by a foam line or a line of detritus.

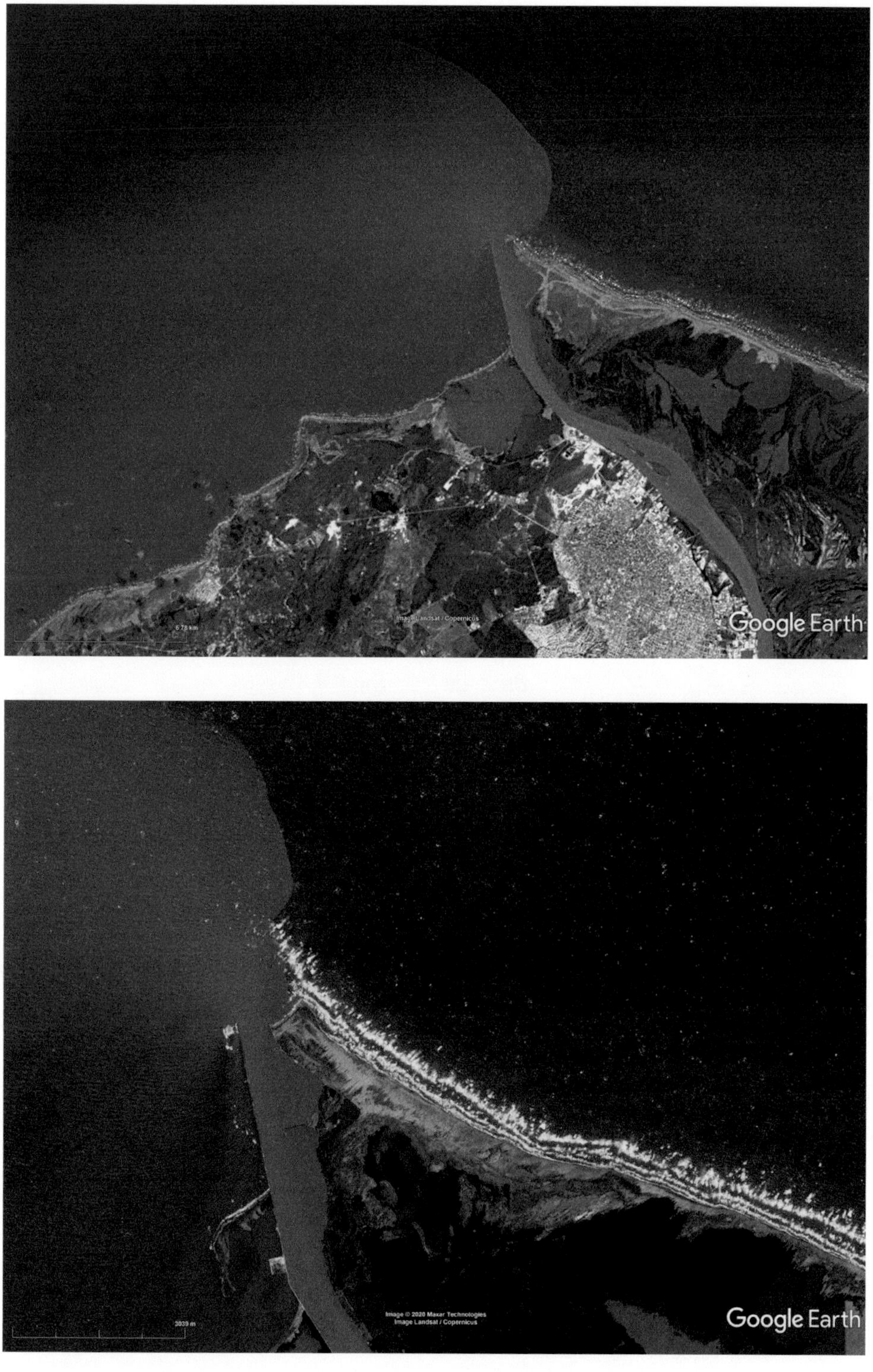

Figure 9.1 Image of a plume front from the Magdalena River, Colombia, discharging onto the Caribbean Sea. Riverine waters in this case are clearly marked by their suspended sediment load.

Plume fronts appear when a buoyant discharge enters ambient waters with different density. The discharge may pulsate with tidal periodicity in such a way that a new plume front is formed with ebbing phases of the tidal cycle. Thus, plume fronts may leave a sequential band of fronts that become less pronounced with distance from shore. Plume fronts are ephemeral and have time scales from minutes to days, depending on the strength of the discharge and the forcing at the receiving waters that will mix the buoyant waters.

At the edge of each plume, there are regions of enhanced gradients that, according to the last term in equation 9.2, have increased convergence rates $\partial v/\partial y$ and/or rapid spatial changes in vertical velocities $\partial w/\partial y$ (Figure 9.2). Examining the same term, $\partial C/\partial z$ is expected to be negative under stable stratification. Hence, a combination of negative $\partial v/\partial y$, that is, flow convergence, and positive $\partial w/\partial y$, that is, vertical downward flow inside the plume (or at the front) greater than that vertical flow at ambient waters, will strengthen horizontal gradients $\partial C/\partial y$. In essence, convergence regions are associated with fronts, while divergence conditions should smear fronts.

Vertical velocities associated with a plume front can be up to several centimeters per second. The magnitude of these velocities can be derived from conservation of mass:

$$\frac{\partial u}{\partial x} + \frac{\partial v}{\partial y} = -\frac{\partial w}{\partial z}. \tag{9.3}$$

Considering only the lateral direction y, we can assume

$$\frac{\partial u}{\partial x} \ll \frac{\partial v}{\partial y}. \tag{9.4}$$

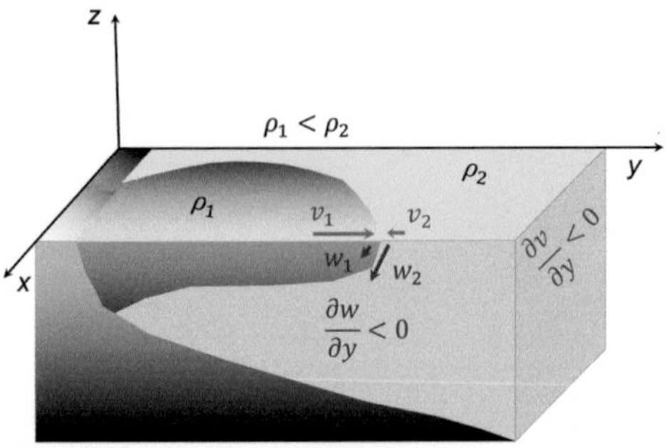

Figure 9.2 Schematic of mechanisms enhancing plume fronts. Low-density water expands offshore. Density-gradient enhancements are favored at the limit of expansion by negative divergence (convergence) and negative horizontal gradients in vertical velocity.

As an example, envision opposing currents 0.40 and −0.10 over a horizontal distance of 10 m in the y direction (e.g., Figure 9.2) and through 2 m of the water column, which are observable conditions. This flow convergence results in

$$\frac{\partial v}{\partial y} \approx \frac{-0.50 \text{ m/s}}{10 \text{ m}} = -0.05\, s^{-1}. \tag{9.5}$$

Rewriting conservation of mass (equation 9.3) with the assumption given by equation 9.4, yields

$$\Rightarrow \frac{\partial v}{\partial y} = -\frac{\partial w}{\partial z}, \tag{9.6}$$

and using the value in equation 9.5 to solve 9.6 results in a vertical downward velocity of 0.1 m/s, associated with a convergent flow:

$$w = -\int_0^{-2} \frac{\partial v}{\partial y} dz \approx -0.10 \text{ m/s}.$$

This order of magnitude (10^{-1} m/s) should be expected for vertical velocities at some frontal zones.

The buoyant discharge that forms plume fronts can have positive (lighter than ambient) or negative (heavier than ambient) buoyancy. Regardless of the sign of the buoyancy discharge, under negligible ambient currents the plume should spread radially away from the source. Most often, however, the fate of the buoyant discharge will be affected by an ambient current or by *Coriolis* accelerations. The ambient current will be driven locally by wind stress or by wind waves or by tides, or remotely by winds, waves, or tides occurring at a different location.

Tidal intrusion fronts are relatives of plume fronts. They are associated with enhanced convergences $\partial v/\partial y$ arising from tidal flows interacting with density gradients. They are also linked to marked horizontal changes in vertical velocities $\partial w/\partial y$ (equation 9.2). Tidal intrusion fronts develop in the early flood stage of tidal flows, typically at or near the entrance to an estuarine system (Figure 9.3). At the end of ebb flows, upon extrusion of relatively low salinity waters out of an estuary, the flood flow encounters near-stagnant low salinity waters and forms a front. At the surface, the collision of salty and heavy flood waters with light estuarine waters causes the heavy waters to plunge underneath the surface waters, marking a well-defined V-shape frontal structure.

The V-shape of the front results in part from the lateral gradients in tidal flow. The apex of the V (Figure 9.3) appears where the inflowing tidal currents are strongest, and where the plunge is most marked. Weaker currents on both sides of the maximum inflow transport heavy water relatively more slowly, resulting in the shape. This process can be visualized in Video 9.1, where the strongest flow (which

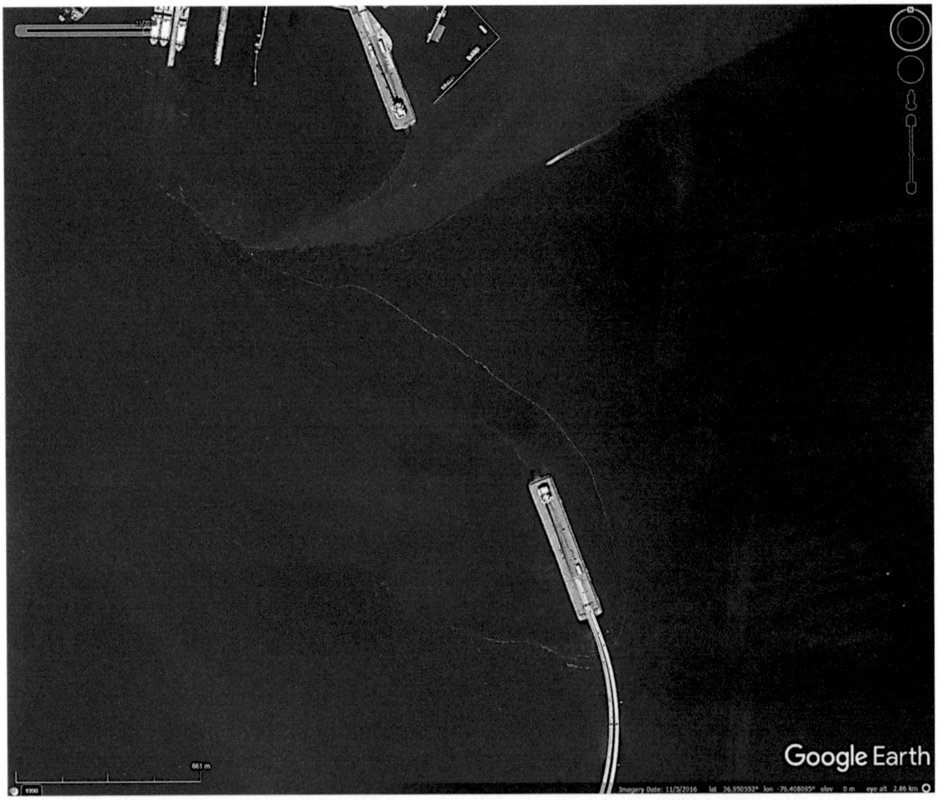

Figure 9.3 Image of a tidal intrusion front off Newport News Point in the James River, Virginia, USA. The image is from November 5, 2016. The tidal intrusion front forms a V-shaped wedge from relatively high salinity water surrounded by relatively low salinity water.

has constant density in the video) is marked by a V-shape convergence. In estuaries, the self-motion of the density contrast contributes to enhancement of the front.

In the water column, the gradient enhancement associated with the tidal intrusion front can be greatest beyond the water surface (Figure 9.4). In the example of Figure 9.4, the largest change of density with time is observed with the onset of flood currents and the greatest vertical stratification just ahead of the maximum inflow. In this tidal intrusion front, the greatest tidal flood currents appear at the pycnocline, that is, well below the water surface, as the baroclinic pressure gradient acts with the tidal flow. Above the pycnocline, the tidal inflow is weakened by the barotropic pressure gradient that tends to drive transport seaward. In basins that feature tidal intrusion fronts, and in systems with a well-defined salt wedge, the pycnocline acts as a slippery boundary where flood currents tend to increase in magnitude. In some basins with a well-defined salt wedge, the maximum flood currents occur at the pycnocline.

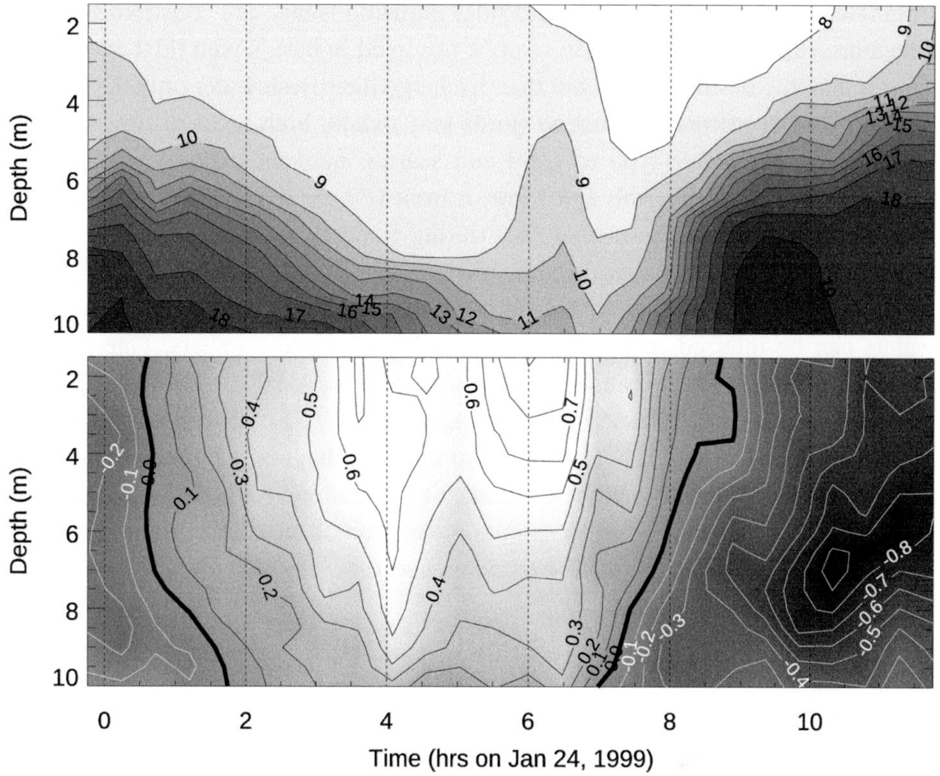

Figure 9.4 Time series of density anomaly (kg/m^3, upper panel) and along-basin flow (m/s, lower panel) during one semidiurnal tidal cycle off Newport News Point, same area as in Figure 9.3. In the lower panel, positive values are seaward (ebb) and the thick black contour is 0 m/s (change of tidal current phase).

In general, tidal intrusion fronts may be described in terms of the influence of inertia (advective momentum flux per unit mass) and baroclinicity for outflowing (positively buoyant) and inflowing (negatively buoyant) layers. Although these layers are rarely homogeneous with depth, we can segregate them above and below a well-defined pycnocline. Tidal intrusion fronts develop when the inertia of the outflowing layer is arrested (zero ebb flow) and heavy water plunges underneath light water. The plunging location may be identified where the tidal inflow speed U_f equals the density-driven flow speed of the incoming layer $(\sqrt{g'H})$, where g' is reduced gravity and H is water column depth (see equation 7.34 in Section 7.3). As U_f increases relative to $\sqrt{g'H}$, the tidal intrusion front propagates into the basin and eventually dampens via mixing.

In most cases, tidal intrusion fronts are constrained by two coastlines roughly oriented in the same direction, whereas plume fronts expand onto ambient waters that are relatively unconstrained. There are cases, however, where both plume-like fronts, which appear during ebb, and tidal intrusion fronts, are constrained inside the same semienclosed basin. This can be observed in basins with tidal excursions shorter than the basin's length and that discharge their fresh water only during ebb phases. Glacial, marine-terminating fjords may exhibit both types of fronts during melting season. In this type of fjord and season, buoyant discharge progresses toward the ocean during ebb and forms a front (Figure 9.5a). Buoyant discharge becomes arrested near the glacier face during flood flows, after a tidal intrusion front appears in early phases (Figure 9.5b).

The dynamics associated with different kinds of plumes expanding on ambient waters can be inferred mainly from two nondimensional numbers. One number compares the offshore expansion of the plume, that is, its offshore scale, to its internal (or baroclinic) Rossby radius (see equation 7.35), known as the Kelvin number (K_e, equation 7.45). The other number that diagnoses plume dynamics is the densimetric Froude number, similar to equation 8.17. It compares the instantaneous current speed in the plume u_p (instead of the tidal current amplitude of equation 8.17) to its expected expansion speed ($\sqrt{g'H}$):

$$Fr^2 = \frac{u_p^2}{g'H}. \tag{9.7}$$

Bulging radial plumes, like those associated with power-plant discharges or with narrow (<1 km wide) inlets, typically have $K_e \ll 1$ and $Fr^2 > 1$. These plumes are "supercritical" relative to $\sqrt{g'H}$, that is, affected by advective accelerations and unaffected by *Coriolis* accelerations. On the other hand, elongated plumes, that is, those that develop into elongated coastal currents, are indeed affected by *Coriolis* and unaffected by advection. In these cases, like the Gaspé current or the Norway, Alaska, Chesapeake Bay, and Delaware Bay coastal currents, $K_e > 1$ and $Fr^2 \ll 1$. As seen, plume shapes may help diagnose their dynamics.

9.3 Mixing Fronts

This type of front occurs at areas of sharp changes in bathymetry, contrasting a vertically mixed water column with a stratified water column. Equation 9.2 indicates that enhanced gradients of any dissolved or suspended material C in seawater can be determined by spatial variations in vertical mixing (stress divergence) $\partial F_z/\partial z$ versus buoyancy S_C. The frictional stress divergence may be scaled in terms of the eddy viscosity and tidal current amplitude as $A_z U_0/H^2$ or in terms of the bottom velocity and drag coefficient as $c_b u_b^2/H$ (e.g., equations 2.42, 4.7, and 5.1). Therefore, we can propose ratios between buoyancy and friction in the following expressions:

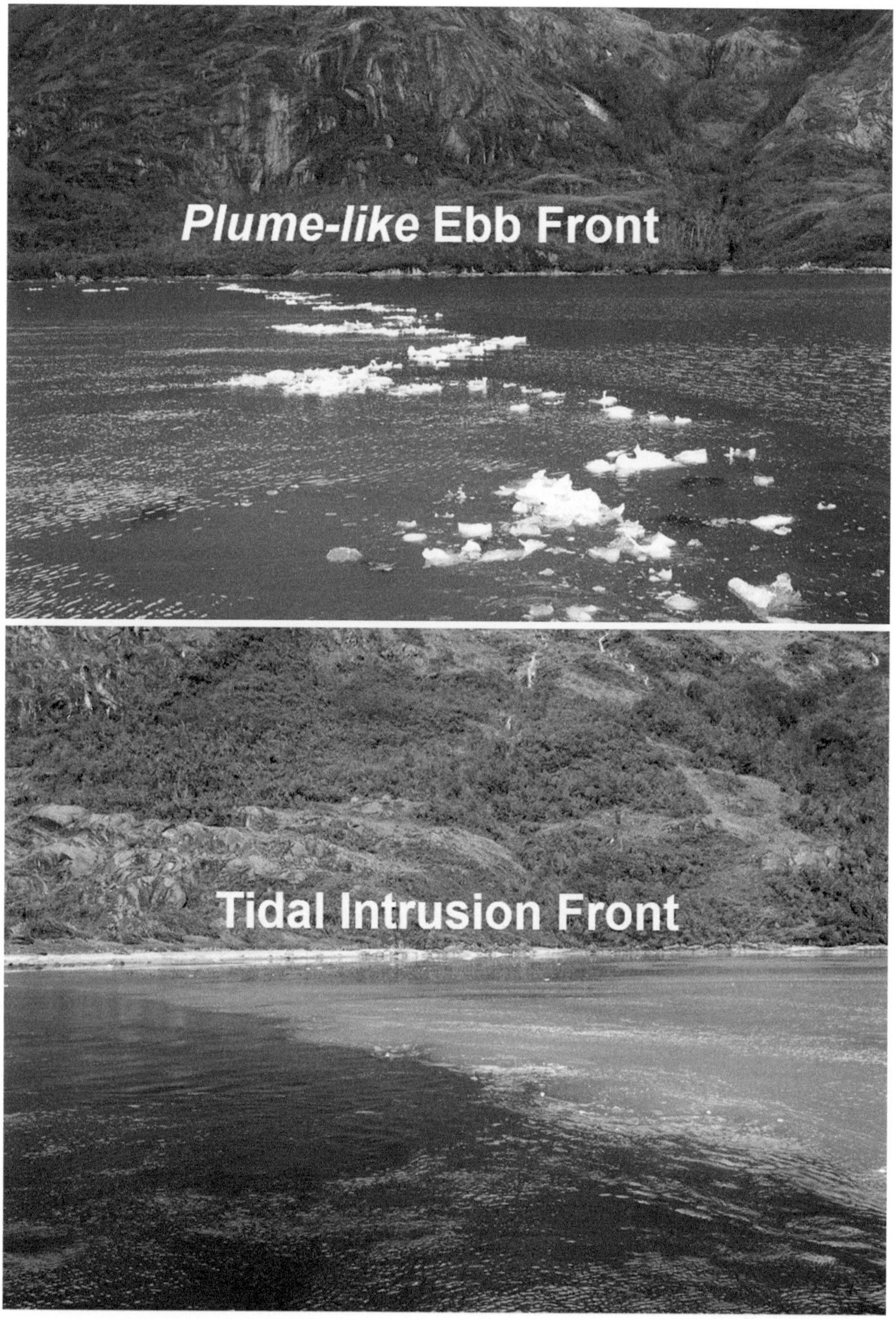

Figure 9.5 Figure of (a) ebb-like and (b) tidal intrusion fronts in Seno Ballena, a glacial fjord off the Strait of Magellan. In (a), the glacier source is to the left and the ocean (Strait of Magellan) is to the right. Ice debris (mini icebergs) and macroalgae fragments are seen accumulating at the front. In (b), ocean water is to the left and glacier (milky) water is on the right. Milky water is only evident in late stages of ebb, not the early stage portrayed in (a).

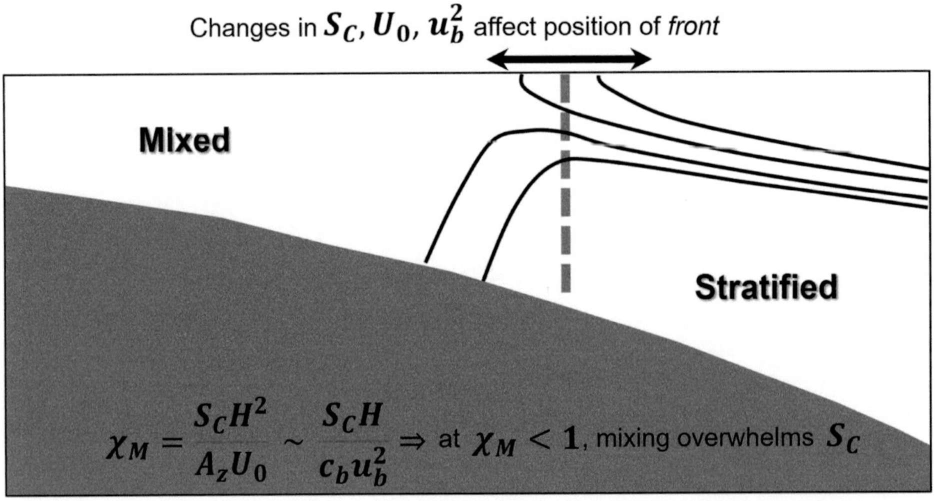

Figure 9.6 Schematic of a tidal mixing front. The front can move offshore, that is, deeper, with increased values of S_C or with reduced values of current strength. Values of S_C may vary from daily to seasonal scales while currents can change from semidiurnal to annual scales.

$$\chi_M = \frac{S_C H^2}{A_z U_0} \sim \frac{S_C H}{c_b u_b^2}. \tag{9.8}$$

The numerator indicates stratification from buoyancy and the denominator accounts for friction by currents. Depending on the buoyancy input to a region, these ratios will establish thresholds above which the water column will remain stratified. Sharp gradients of these ratios will likely indicate locations of *tidal mixing* fronts (Figure 9.6). Furthermore, the position of the front will change with variations of S_C and U_0 (or u_b^2) at time scales from tidal to seasonal. Buoyancy input S_C can develop from freshwater discharge or heating. In turn, u_b^2 may be produced by tides but also by wind-driven currents.

9.4 Shear Fronts

The third (or fourth, depending on whether plume and tidal intrusion fronts are treated differently) general type of front is the *shear* or *lateral* or *axial* front. It is probably the most easily and most frequently observed type of front inside estuaries. Shear fronts result from the differential advection by tidal currents $\partial u/\partial y$ of an along-basin gradient in property C, or, $\partial C/\partial x$, or writing the relevant terms in equation 9.2:

$$\frac{\partial}{\partial t}\left\{\frac{\partial C}{\partial y}\right\} = -\frac{\partial u}{\partial y}\frac{\partial C}{\partial x}. \tag{9.9}$$

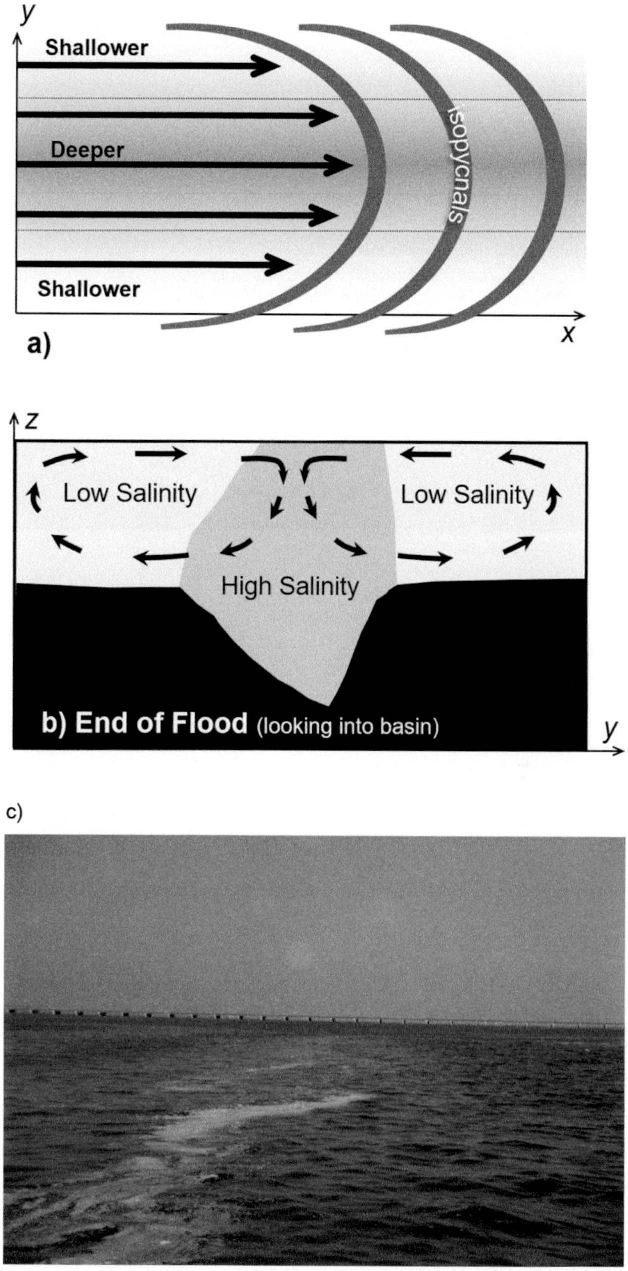

Figure 9.7 Schematic of shear fronts (a) with corresponding lateral flow structure with depth (b). Shown also are photographs of along-estuary fronts during ebb (c) and flood (d) phases of the tidal cycle.

d)

Figure 9.7 (*cont.*)

Over a typical channel-shoals bathymetry, tidal flows are stronger in the channel than over shoals (Section 3.4.6). The lateral distribution of tidal flows, that is, the lateral shears in tidal flows, strains the water density (or salinity, or property *C*) field laterally. During the flood phase of the tidal cycle, the lateral straining causes higher density water to appear in the channel, relative to shoals (Figure 9.7a). Along-basin fronts will develop by flow convergence caused by two processes: (i) lateral shears at scales of order meter and (ii) lateral density gradients.

Convergences linked to lateral shears in along-basin flows can appear under homogeneous fluids, as shown in Video 9.2, in river flow. The video illustrates eddies with diameters <1 m at the transition from relatively weak currents to relatively strong currents. A line of convergence can be appreciated along the transition, which is represented by maximum lateral shears in the field of view. Through vorticity considerations that go beyond the scope of this introductory text, we can show that convergences are linked to lateral shears in along-basin flows, as shown qualitatively in the video.

Convergences induced by lateral density gradients may drive along-estuary or axial fronts. Lateral or secondary flows display confronting motion from shoals to channel (convergence) at the surface and motion away from the channel (divergence) at depth (Figure 9.7b). Following the same reasoning, no axial front would develop in ebb phases. Lateral straining from laterally sheared tidal flows during ebb would cause lower density water in the channel relative to the shoals. These lateral gradients would therefore trigger near-surface divergence of lateral flows and convergence at depth. However, shear fronts can develop during ebb phases of the tide, too.

Depending on the width of the channel-shoal systems, and for reasons yet to be clarified, shear fronts can appear in the middle of the channel or at the edge of a channel (Figure 9.7c and d illustrating fronts at edges). Along-estuary fronts may develop on the right edge (in the northern hemisphere) relative to tidal flow direction, regardless of the tidal phase. Reasons for along-estuary fronts appearing over the right flank of a channel (in the Northern Hemisphere) have to do with the phase lag of tidal currents between channel and shoals (Section 3.4.6).

Considering that a tidal wave entering a semienclosed basin would follow a cyclonic rotation (as in a Kelvin wave around a basin; Section 3.4.4), phase changes will occur over shoals before the channel. As tidal currents rotate cyclonically, at the end of flood the currents over the shoals will point to the left (right in the Southern Hemisphere) of the current in the channel. This will cause convergence over the right edge of the channel and divergence on the left (Figure 9.8a). In the Southern Hemisphere, convergences are expected on the left edge of the channel (looking into the basin) at the end of flood for the same reasons. At the end of ebb, the phase advance of the tidal currents over shoals, relative to the channel, will cause flows directed to the left of the ebb flow in the channel (Figure 9.8b). This would cause convergence over the left edge, looking into the basin, that is, to the right of the tidal flow direction in the channel. Divergence would appear on the other edge. Convergence locations would be opposite in the Southern Hemisphere, as Kelvin waves rotate in the opposite direction.

A similar argument that explains near-surface convergences on the edge of a channel to the right of the tidal flow direction can be as follows. Toward the end of each tidal phase, tidal currents over shoals already flow in opposite direction to those in the channel. In the Northern Hemisphere, convergence of lateral flows will appear to the right of the tidal flow in the channel as *Coriolis* act on these opposing flows (Figure 9.8).

Appearance of near-surface convergence at the edge of a channel to the right of the tidal flow in the channel is illustrated with observations at the entrance to Chesapeake Bay. Four tidal cycles of surface velocity measurements and of estimates of lateral divergence $\partial v / \partial y$ consistently show negative divergence on the channel edge to the right of the tidal flow direction (Figure 9.9). There is still much work to investigate the influence of bathymetry and density fields on the generation and persistence of along-basin fronts. These features tend to accumulate materials and micro-organisms that, despite their ephemeral existence, have crucial ecological impact.

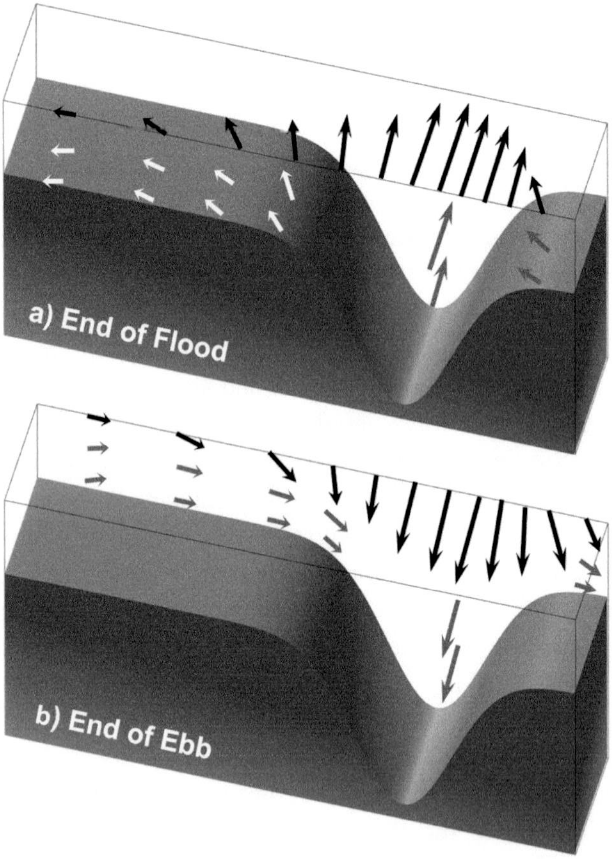

Figure 9.8 Schematic of convergence process at channel edges. View is into the basin in the Northern Hemisphere. In the Southern Hemisphere, the view would be seaward and the labels would change between "flood" and "ebb." Darkest arrows describe near-surface flow and non-black arrows indicate flows in the water column.

9.5 Take-Home Message

Fronts are easy to see but difficult to define and study because of their ephemeral nature. We can study fronts using a conservation equation. This approach identifies plume, mixing, and shear fronts. The general shape of plume fronts hints at their dominant dynamics, which reflect a densimetric Froude number: radial plumes are inertial, and elongated plumes are affected by *Coriolis* accelerations. Tidal intrusion fronts may be characterized also with a densimetric Froude number. Mixing fronts tend to occur at the edge of sharp bathymetric changes, as well as shear fronts. Mixing fronts are caused by diffusive effects, while shear fronts are linked to advective agents.

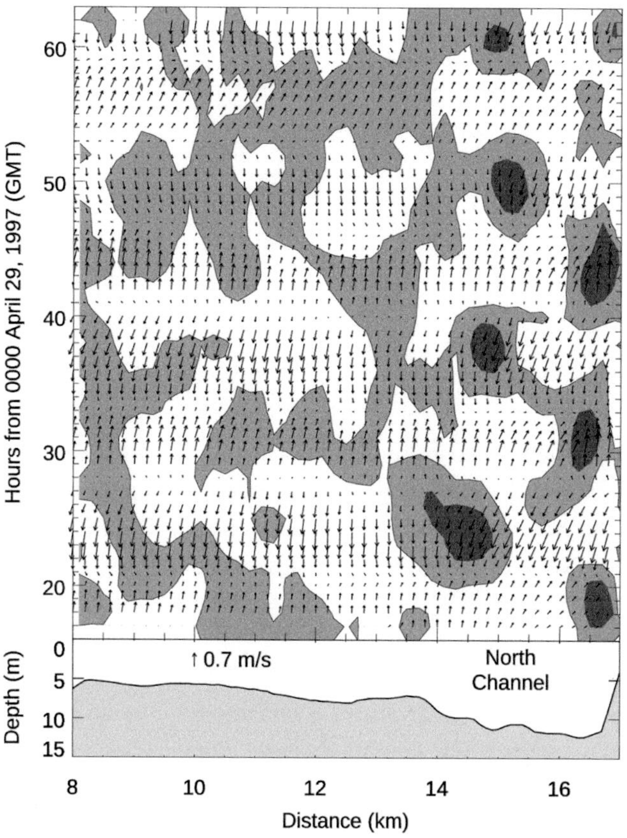

Figure 9.9 Phase diagram (distance vs. time) of near-surface flows and lateral convergence $\partial v/\partial y$ at the edge of the North Channel, at the entrance to Chesapeake Bay. Vectors indicate surface flows, with flood flows pointing upward on the plot. Contours are values of $\partial v/\partial y$, with convergences shaded in gray. Darkest gray shades denote strongest convergence values. Underneath the phase diagram is the bathymetry of the transect.

Additional Sources

Garvine, R.W. (1995) A dynamical system for classifying buoyant coastal discharges. *Continental Shelf Research* 15(13): 1585–1596.

Horner-Devine, A.R., R.D. Hetland, and D.G. MacDonald (2015) Mixing and transport in coastal river plumes. *Ann. Rev. Fluid Mech.* 47(1): 569–594.

O'Donnell, J. (1993) Surface fronts in estuaries: A review. *Estuaries* 16: 12–39.

O'Donnell, J. (2010) The dynamics of estuary plumes and fronts. In *Contemporary Issues in Estuarine Physics*. Edited by A. Valle-Levinson, pp. 186–246. Cambridge: Cambridge University Press.

10

Time Scales in Semienclosed Basins

Studies of any semienclosed basin frequently ask the question, How long does it take for water renewal in this basin? The question is rather simple, but the answer is far from it. The response is difficult because one single value (e.g., N_T number of hours or days or years) falls short in representing the possible variability of the agents that cause such water renewal. This chapter presents different ways to represent the renewal and the potential variability in those representations. The chapter is anchored in previous syntheses that attempt to address persistent inconsistencies in the use of the various terms that refer to such water renewal. The chapter proposes distinctions among *flushing time*, *residence time*, *age*, *transit time*, and *exposure time*.

10.1 Flushing Time

Flushing time, also referred to as *turnover time*, can actually be defined from two perspectives: (a) the influence by the residual (or long-term mean) volume flux (in m^3/s) from the ocean F_{in} and (b) the influence by the river discharge R (m^3/s).

10.1.1 Flushing Time by Ocean Volume Flux F_{in}

In this approach, we concentrate on the end of the basin forced by the ocean. Considering F_{in} flushing time (T_{fF}) is the time it takes to replace the entire volume of a semienclosed basin V by the residual or non-tidal volume influx F_{in} (Figure 10.1a):

$$T_{fF} = \frac{V}{F_{in}}, \tag{10.1}$$

where T_{fF} is given in seconds. The key in this approach is to know F_{in}, which can be derived from the combination of steady-state (i.e., time invariant)

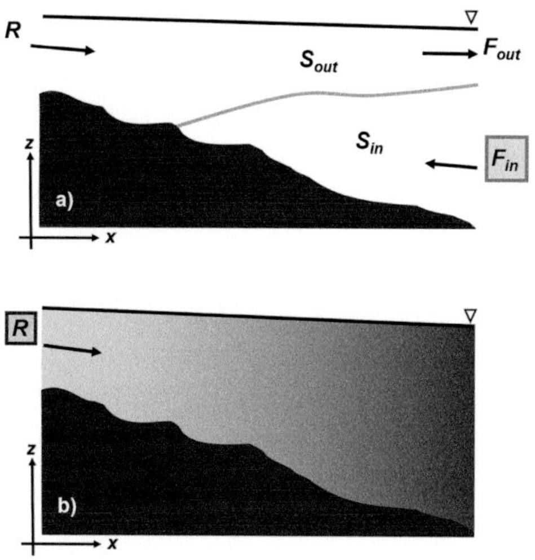

Figure 10.1 Schematic of a longitudinal section along a basin for the framework related to flushing time by residual volume influx F_{in} and by river discharge R. (a) Flushing time depends on F_{in}, which is calculated from Knudsen's equations (10.2 to 10.4). (b) Flushing time depends on the river discharge R according to equations 10.5 or 10.8.

basin-integrated conservation of mass (water) and conservation of salt (Chapter 2), also known as *Knudsen's* or *basin's equations* (refer to Figure 10.1a):

$$\text{Water budget}: \quad F_{out} = F_{in} + R \tag{10.2}$$

$$\text{Salt budget}: \quad F_{out} S_{out} = F_{in} S_{in}. \tag{10.3}$$

In equations 10.2 and 10.3, F_{out} is the residual volume flux (in m^3/s) out of the basin, while S_{out} and S_{in} are the residual, or non-tidal, salinities of outgoing and incoming waters, respectively. These equations (10.2 and 10.3) represent a system of two algebraic equations that can be solved for F_{out} and F_{in}, knowing R, S_{out} and S_{in}. Solving the system of equations (10.2 and 10.3) yields

$$F_{in} = R \frac{S_{out}}{(S_{in} - S_{out})}. \tag{10.4}$$

The result in equation (10.4) is then used in equation 10.1 to obtain the flushing time by the volume flux into the basin. It is evident that F_{in}, and hence T_{fF}, is sensitive to the difference in salinities between residual outflowing and inflowing waters. Such difference changes at various temporal scales. This flushing time estimate is also responsive to R, which may vary synoptically, monthly, seasonally, and at longer time scales. Therefore, any semienclosed basin will display a range of possible T_{fF} values, not only one.

10.1.2 Flushing Time by River Flux R

In this alternative approach to determine *turnover* times, we consider the end of the basin forced by the river discharge R or river flux over an entire cross-section. Flushing time by the river T_{fR} (in seconds) is then the time required to replace the freshwater volume in the system V_f (in m^3) by the river discharge (Figure 10.1b):

$$T_{fR} = \frac{V_f}{R}. \tag{10.5}$$

Once again, it is clear that this estimate of *flushing time* is sensitive to the variability of R, which can exhibit a gamut of values. Estimates in equation 10.5 also respond to the freshwater volume V_f, which varies widely. A major challenge of this approach is precisely the evaluation of V_f.

One way to determine V_f is with the concept of freshwater fraction f_r, which can be defined in terms of the oceanic salinity S_0 and the salinity throughout the basin $S(x, y, z)$, as

$$f_r = \frac{S_0 - S}{S_0}. \tag{10.6}$$

A value of f_r equals zero indicates that S has oceanic salinity, that is, no freshwater fraction or presence. On the other hand, a value of f_r equals one represents zero salinity ($S = 0$), that is, only freshwater. Thus, the volume of freshwater for equation 10.5 may be obtained by integrating the three-dimensional freshwater fraction over a basin, that is,

$$V_f = \int_V f_r dV. \tag{10.7}$$

In an ideal situation in which we have measurements or numerical model results of salinity profiles, and values of f_r, throughout a basin, we use equation 10.7 directly. However, in more practical circumstances, we can approximate f_r with a spatially averaged value $\overline{f_r}$, in such a way that $V_f = \overline{f_r} V$, and therefore

$$\bar{T}_{fR} = \frac{\overline{f_r} V}{R}. \tag{10.8}$$

In the simplest scheme, values of $\overline{f_r}$ may be calculated as an average of the salinity value at the head and mouth of the basin. As seen, flushing times given by equations 10.1, 10.5, and 10.8 neglect the effects of tidal flows. Tidal influence is explored in the following subsection.

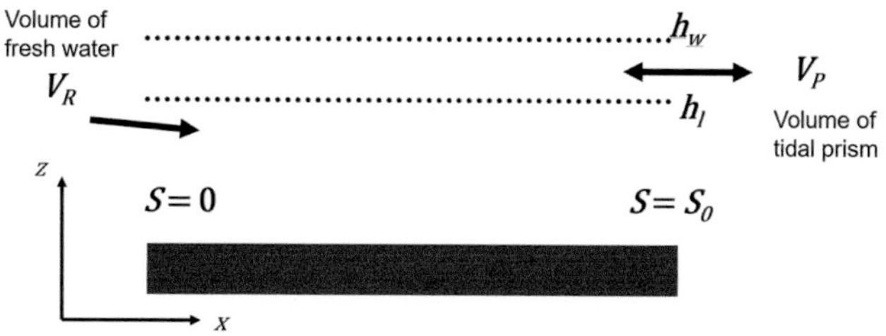

Figure 10.2 Schematic of an along-basin section illustrating the tidal prism (equation 10.9) and the salinities at the seaward and landward ends of the basin.

10.1.3 Flushing Time by Tides

An additional method to establish flushing times considers the influence of tides. This is sometimes known as the *tidal prism method*. So first we need to understand that *tidal prism* (V_P) is the volume of water that enters and leaves a semienclosed basin with every tidal cycle. It may be calculated as the product of the tidal range times the horizontal, or surface, area A_S of the basin, that is,

$$V_P = \underbrace{(h_w - h_l)}_{\textit{tidal range}} A_S, \tag{10.9}$$

where h_w is high tide and h_l is low tide (Figure 10.2).

To determine flushing time with the tidal prism method, we apply five restrictive assumptions. First, we assume a "short" basin in which there are no variations in depth. Second, we suppose that the volume of freshwater that enters the basin throughout one tidal cycle V_R is mixed completely with V_P. Third, we presume that the volume of mixed water is removed entirely from the basin in the ebb phase. Fourth, the mixing is repeated in the next flood phase with seawater of salinity S_O entering the basin. Fifth, water leaving the basin has a mixed salinity $\breve{S}$.

Values of $\breve{S}$ at high tide are calculated from the basin-integrated salt volume at steady state

$$V_P S_0 = [V_P + V_R]\, \breve{S}, \tag{10.10}$$

which yields, solving for $\breve{S}$,

$$\breve{S} = \frac{V_P}{V_P + V_R} S_0. \tag{10.11}$$

Equation 10.11 indicates that if $V_P \gg V_R$, that is, if the tidal prism is much greater than the volume of freshwater, then the mean salinity tends to the oceanic salinity $\breve{S} \to S_0$. The equation also implies that if $V_P \ll V_R$, then $\breve{S} \to 0$, that is, the basin becomes dominated by freshwater. Then, the average or mean freshwater fraction (defined right before equation 10.8, but now considering the tidal prism) becomes

$$\overline{f_{rT}} = S_0 - \frac{\breve{S}}{S_0} = 1 - \frac{\breve{S}}{S_0} = 1 - \frac{V_p}{V_p + V_R} = \frac{V_R}{V_p + V_R}, \tag{10.12}$$

which uses equation 10.11 for $\breve{S}$. This expression for average freshwater fraction includes the effects of river V_R and tides V_P.

Taking now the flushing time that considers the average freshwater fraction (equation 10.8):

$$T_{fT} = \frac{\overline{f_{rT}}V}{R} = \frac{\overline{f_{rT}}V}{V_R/T}, \tag{10.13}$$

where T is a period, taken here as the *tidal period* (in seconds). Then, inserting the value of $\overline{f_{rT}}$ (equation 10.12) in equation 10.13, after a slight manipulation produces

$$T_{fT} = \frac{VT}{V_p + V_R}. \tag{10.14}$$

This flushing time now considers tidal effects, through the tidal period and prism. One implication drawn from equation 10.14 is that if the volume of the basin V is much greater than the tidal prism and the freshwater volume, that is, if $V \gg V_P + V_R$, then $\bar{T}_{fT}/T \gg 1$. In other words, it would take many tidal cycles T to flush the basin. Conversely, it would only take a fraction of a tidal cycle to flush the basin if $V < V_P + V_R$.

One drawback of this approach to estimate flushing time is that it assumes basin-wide mixing of waters entering the system. This assumption provides underestimations of flushing times relative to the actual values because $\breve{S}$ will likely be smaller than the value in equation 10.11. Another shortcoming is that the approach considers no atmospheric forcing and that the water entering the basin has oceanic salinity. Usually, the water entering back into a basin is a mixture of oceanic and basin's waters, at least early in the tidal cycle. Thus, values of flushing time from equation 10.14 will be shorter than those obtained from equations 10.1, 10.5, and 10.8. One way of lessening these inadequacies is to subdivide the basin in segments and calculate the *flushing rate* at each segment. To do that, we first introduce the concept of flushing rate.

10.1.4 Flushing Rates

Flushing rate F (in m^3/s) is the rate at which the entire volume of a basin V is exchanged or renewed. It is roughly the inverse of the flushing time. Using the river flushing time T_{fR},

$$F = \frac{V}{T_{fR}}. \tag{10.15}$$

Using equation 10.5 for the denominator in equation 10.15, and the finding that $V_f = \overline{f_r} V$, provides

$$F = \frac{VR}{V_f} = \frac{R}{\overline{f_r}}. \tag{10.16}$$

Following the same arguments, the flushing rate caused by the tidal prism F_T becomes

$$F_T = \frac{V}{T_{fT}} = \frac{V_p + V_R}{T}. \tag{10.17}$$

Intuitively, flushing rates by tides should be greater than those by river discharge R or by residual volume inflow F_{in}. In other words, tides typically exchange waters more readily than residual flows. These concepts should then be used cautiously, depending on the temporal scales of interest.

10.1.5 Flushing Rates in Segments of a Basin

As mentioned at the end of Section 10.1.3, one way to reduce the disadvantages of the tidal prism method is to partition a basin in segments. We assume that there is complete mixing at each segment in each tidal cycle rather than throughout the entire basin.

The length of each basin's segment is defined by the tidal excursion (L_T in meters), as $U_0 T/\pi$ (see equation 8.15), where U_0 is the tidal current amplitude. In turn, all i basin segments have an intertidal volume P_i and a low water volume VL_i (Figure 10.3a). The intertidal volume for each segment contains the tidal prism V_{Pi} and the freshwater volume V_{Ri}, that is,

$$P_i = V_{Pi} + V_{Ri}. \tag{10.18}$$

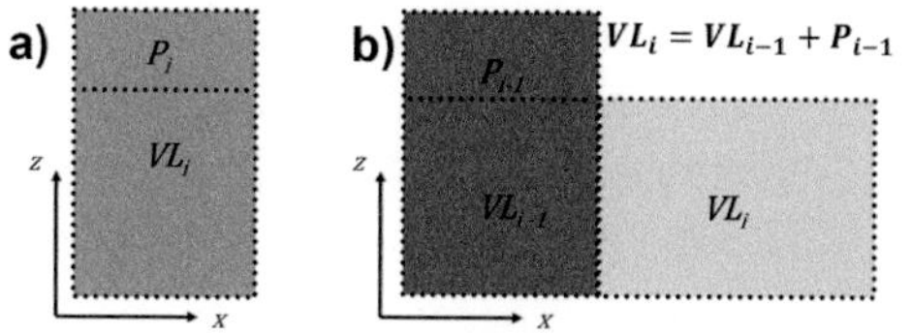

Figure 10.3 Longitudinal ith segment in a basin showing (a) its intertidal volume P_i and its low-tide volume VL_i and (b) the geometric relationship between contiguous segments in the basin.

At the landward end of the basin, the low tide volume $VL_i = VL_0$, and the intertidal volume can be defined as $P_i = P_0 = V_{R0}$, that is, there is no tidal variation and the volume variations in the segment are only produced by river-water volume. For the segments that follow seaward from the landward end, assume that the low-tide volume of the seaward segment equals the high-tide volume of the landward segment (Figure 10.3b), that is,

$$VL_i = VL_{i-1} + P_{i-1}. \tag{10.19}$$

This implies either that the tidal excursion decreases landward or that the basin becomes narrower, or both. In general, the volume of all segments is given by

$$\begin{aligned}
VL_1 &= VL_0 + P_0 \\
VL_2 &= VL_1 + P_1 = VL_0 + P_0 + P_1 \\
VL_3 &= VL_2 + P_2 = VL_0 + P_0 + P_1 + \mathrm{P}_2 \\
&\vdots \\
VL_i &= VL_{i-1} + P_{i-1} = VL_0 + \sum_{m=0}^{i-1} P_m = VL_0 + V_{R0} + \sum_{m=1}^{i-1} P_m
\end{aligned} \quad . \tag{10.20}$$

Complete mixing in each segment at high tide implies that the proportion of water removed from the segment during the ebbing tide is given by the ratio between intertidal volume P_i and the segment's high-tide volume $(VL_i + P_i)$, that is,

$$r_i = \frac{P_i}{VL_i + P_i}. \tag{10.21}$$

The variable r_i is the *exchange ratio* of segment i, which is nondimensional with values between 0 and 1. If VL_i equals zero, it means that the segment goes dry at low tide and r_i becomes 1. This indicates complete exchange of waters in the segment, as on a tidal flat. On the other hand, the *exchange ratio* will tend to zero as the intertidal volume P_i decreases relative to the low-tide volume VL_i and approaches zero. We can thus, calculate r_i for segments along a basin and obtain a spatial distribution of *exchange ratios* throughout a basin.

To attain a spatial distribution of flushing times throughout segments of a basin, we can use equation 10.14 and note that the volume of the basin equals the low-tide volume plus the intertidal volume ($V = VL_i + P_i$). Also note that P_i is a modified tidal prism that includes the tidal prism volume V_p plus the freshwater volume V_R, which will be negligible as we move seaward. Thus, from equation 10.14

$$\frac{V}{V_p + V_R} = \frac{VL_i + P_i}{P_i} = \frac{1}{r_i}.$$

Consequently, the tidal flushing time for each segment i of a basin can be proposed as:

$$T_{fTi} = \frac{T}{r_i}. \tag{10.22}$$

This expression provides spatially variable flushing times in a basin. Values will represent multiples or fractions of tidal periods T over which waters are renewed in a segment. Such an approach is likely more enlightening than values derived from equations 10.1, 10.5, 10.8, and 10.14. Furthermore, with this approach we can determine the accumulated freshwater volume per segment:

$$V_{fi} = T_{fTi} R = \frac{T}{r_i}\frac{V_R}{T} = \frac{V_R}{r_n}. \tag{10.23}$$

Finally, we can also calculate the mean salinity of each segment:

$$\breve{S}_i = \frac{V_{fi}}{V_i} S_0 = \left(1 - \overline{f_{rTi}}\right) \cdot S_0.$$

The same reasoning as for salinity can be applied to other dissolved or suspended materials. Keep in mind that these estimates also necessarily have rich variability in time. Therefore, one distribution of T_{fTi} along a basin is insufficient to represent it properly.

Increasing the amount and reliability of information on times for water or material renewal requires expanding the level of complexity. The rest of the concepts articulated in this chapter involve implementation of particle-tracking algorithms in numerical model simulations. The following concepts provide information that depends on time and space and therefore are typically represented as maps or cross-sections. These maps ultimately illustrate the capacity of a basin to exchange its waters with the adjacent ocean.

10.2 Residence Time

Residence time is the time it takes for water (particles or parcels) or material elements to exit a basin. These water or material elements are located initially at time t_0, at any position in a basin. Therefore, this concept is space-dependent as elements found further away from the entrance will take longer to exit.

Residence time is counted at the element's first exit from the basin as it could reenter in subsequent tidal cycles or during synoptic pulses (Figure 10.4). In Figure 10.4a, residence time is obtained as the difference $t_1 - t_0$, where t_1 is the time at which the element exits the basin. This time difference, between the time at which we begin to examine the element and its exit time, is related only to one element or particle. Different elements located at different positions in the basin will obviously have distinct residence times. That is how we can construct a map of

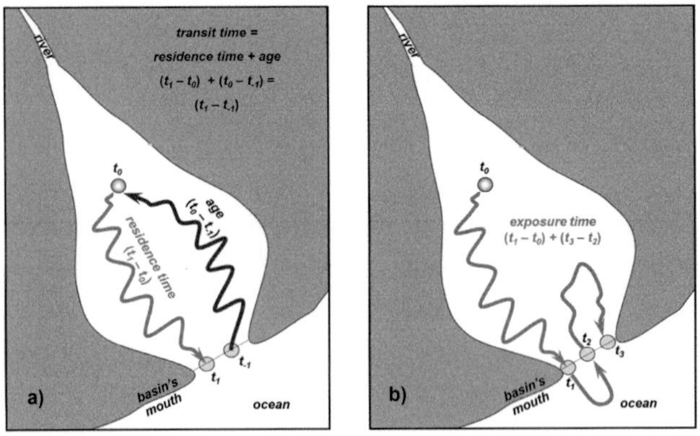

Figure 10.4 Illustration of various concepts related to exchange of properties between a basin and its adjacent ocean. The concepts are exemplified by the hypothetical trajectory of one particle inside the basin considered at time t_0. Time t_{-1} is when the particle entered the basin, t_1 is when it will exit, t_2 (in (b)) is when it will reenter, and t_3 (also in (b)) is when it will exit again.

residence time based on the position of each element. Say, at (x_1, y_1) the residence time is T_1, at (x_2, y_2) the residence time will be T_2, and so on. Again, such a map will apply to a given depth and to certain set of forcing conditions on the basin, such as river discharge, tidal forcing, wind forcing, and, in some cases, wind-wave forcing.

Extensive representations of residence time can be advanced further by considering the initial depth of a particle and its depth at exit. This approach augments complexity because the particle needs prescription of a behavior, that is, neutrally buoyant, migrating vertically or reacting to different salinities. Therefore, a complete description of residence time shall be three dimensional and responsive to distinct forcings.

10.3 Age

Age is the time that a water particle, parcel, or material element has spent inside the basin since its entrance (Figure 10.4a). Age is calculated with the expression $t_0 - t_{-1}$, where t_{-1} is the time at which the element entered the basin. If particles or elements follow the same path into the basin as that out of the basin, residence time and age maps will be the same. However, this overlap is unlikely to occur in natural systems and the maps will be different, thus providing complementary information. This metric has become widely used as an indicator of the ability of a system to flush. Its preferred use is possibly related to its straightforward calculation from numerical model simulations. In contrast to residence time calculations, age estimates prescind waiting until particles exit the basin in the model simulations (e.g., Figure 10.5). Similarly to residence time, age is a three-dimensional metric.

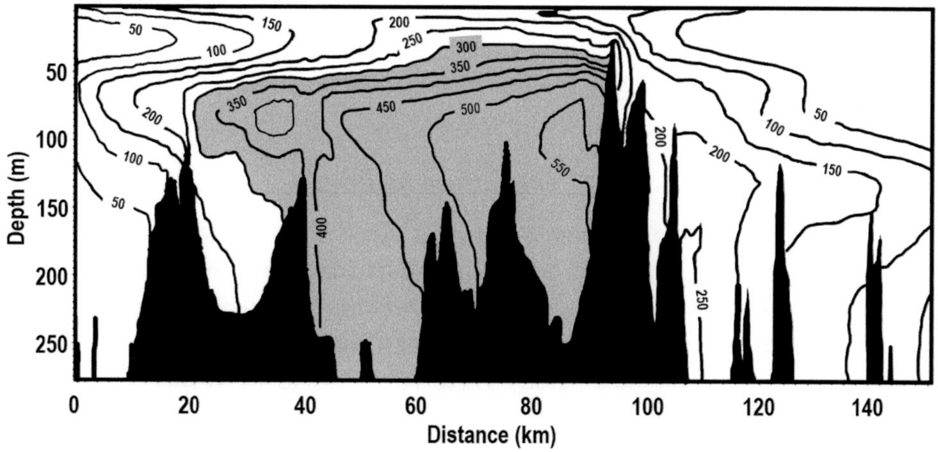

Figure 10.5 Longitudinal section of water age (days) in a Chilean fjord, derived from numerical simulations. Connections to the ocean are to the left and right, and the contours represent mean age over a two-year period. This portrayal clearly illustrates the portion of the basin that has retained water for the longest (shaded region >300 d, i.e., oldest water).
Results provided by Elías Pinilla

10.4 Transit Time

Transit time is the time it takes a particle, parcel, or material element to remain in a basin since the time it entered. It is the same as the sum of age plus residence time, that is, $t_1 - t_{-1}$, or time of entrance minus time of exit. This metric is also three dimensional and should provide larger values than either age or residence time.

The concepts of residence time, age, and transit time can be related to the metric used in living organisms, in particular humans. Taking our birthday as the time we enter the world, such a date is analogous to the time at which a particle enters a basin. Then, our age is how long we have been in the world (alive), analogous to a particle since entering a basin. Our residence time is the time we have left in our lives, counting from now. This is analogous to the time a particle takes to exit the basin. Finally, our transit time is the time from our birth until our death, which is analogous to the time a particle remains inside the basin from its entrance to its exit.

10.5 Exposure Time

Exposure time considers situations in which water particles or material elements reenter a basin after departing from it (Figure 10.4b). This metric includes the residence time of an element and its transit time(s) after it reenters the basin. Using the analogy of human life, exposure time would be the time between now and "temporary death" (after a successful resuscitation), plus the time alive after being resuscitated. This concept is quite useful for pollutants as a basin is susceptible to exposure times rather than residence times. This three-dimensional concept should be used widely when studying the fate of materials and transport of larvae.

10.6 Take-Home Message

We began the chapter with a simple question: What is the renewal time for water, or any dissolved or suspended material, in a basin? The answer is necessarily complicated. There is no single quantity that provides a reliable metric. Rather, a range of values plus a series of maps and/or contours give us an idea of the capacity of a system to flush itself. There is a need to be consistent with the concepts of flushing time, residence time, age, transit time, and exposure time.

Additional Sources

Bolin, B., and H. Rodhe (1973) A note on the concepts of age distribution and transit time in natural reservoirs. *Tellus* 25: 58–62.

Deleersnijder, E., J.M. Campin, and E.J.M. Delhez (2001) The concept of age in marine modelling I. Theory and preliminary model results. *J. Mar. Syst.* 28: 229–267.

Delhez, E.J.M. (2013) On the concept of exposure time. *Cont. Shelf Res.* 71: 27–36.

Delhez, E.J.M., J.M. Campin, A.C. Hirst, and E. Deleersnijder (1999) Toward a general theory of the age in ocean modelling. *Ocean Model.* 1: 17–27.

Delhez, E.J.M., A.W. Heemink, and E. Deleersnijder (2004) Residence time in a semi-enclosed domain from the solution of an adjoint problem. *Estuar. Coast. Shelf Sci.* 61: 691–702.

Lucas, L.V. (2010) Implications of estuarine transport for water quality. In *Contemporary Issues in Estuarine Physics*. Edited by A. Valle-Levinson, pp. 273–307. Cambridge: Cambridge University Press.

Lucas, L.V., and E. Deleersnijder (2020) Timescale methods for simplifying, understanding and modeling biophysical and water quality processes in coastal aquatic ecosystems: A review. *Water* 12(10): 2717.

Monsen, N.E., J.E. Cloern, L.V. Lucas, and S.G. Monismith (2002) A comment on the use of flushing time, residence time, and age as transport time scales. *Limnol. Oceanogr.* 47: 1545–1553.

Takeoka, H. (1984) Fundamental concepts of exchange and transport time scales in a coastal sea. *Cont. Shelf Res.* 3: 311–326.

Zimmerman, J.T.F. (1976) Mixing and flushing of tidal embayments in the western Dutch Wadden Sea. Part I: Distribution of salinity and calculation of mixing time scales. *Neth. J. Sea Res.* 10: 149–191.

11

Semienclosed Basins with Low or No Discharge

This chapter describes semienclosed basins that receive freshwater input in relatively low volumes or with highly seasonal periodicity. In temperate basins, density gradients are dominated by salinity. In contrast, in basins with low or no freshwater discharge (described in this chapter), density gradients may also or exclusively be dominated by thermal gradients. Therefore, research on these basins, which has been scantily reported in the literature, should consider cooling and evaporation. In general, studies in these basins should include assessment of heat and water exchanges with the atmosphere, in addition to land-derived freshwater input. We know less about these systems than those in temperate latitudes because they are most likely found in tropical or subtropical latitudes or near the poles, where access to study sites is less frequent.

We begin the material of the chapter with definition of terms that require careful distinction. Then the framework is established to study these systems in terms of the sign of the temperature and salinity gradients within the system. Such framework enables differentiation among semienclosed basins as "typical" or ***inverse***. Within typical and inverse estuaries, the chapter discriminates among (a) typical but thermally driven, (b) inverse but thermally driven, (c) hyposaline but not typical, and (d) hypersaline but not inverse.

11.1 Definitions

First, a typical estuary or typical semienclosed basin is that in which the residual circulation in the basin consists of outflow of buoyant waters and inflow of relatively heavier waters (as in Chapters 7 and 8). In this case, the density gradient is defined as positive ($\Delta\rho > 0$), that is, water density increasing seaward from inside the basin. An inverse semienclosed basin has to do with its residual circulation, too. In an inverse basin, outflowing waters are relatively heavier and inflowing waters transport relatively lighter waters. In an inverse basin, density gradients are such that $\Delta\rho < 0$, that is, basin waters are denser than in the adjacent coastal ocean.

A *negative* basin or estuary exhibits a negative water balance with more water losses than gains. It implies a negative hydrological balance in which evaporation exceeds precipitation plus river discharge. A negative basin thus displays net buoyancy loss. Negative basins may behave as *typical* but most likely display *inverse* character.

In a typical temperate estuary, the basin is *hyposaline*, meaning that the basin's waters have lower salinity than the adjacent ocean. By convention, these hyposaline systems display $\Delta S > 0$, that is, salinity increasing seaward from inside the basin. A basin can also be *hypersaline* when the basin's waters have greater salinity than the adjacent waters. This *hypersalinity* will be caused most often by exceeding evaporation over precipitation plus river input. Hypersalinity can also result from transport of buoyant water on the coastal ocean to the mouth of a coastal lagoon, or from freezing in part of the basin with brine release. In hypersaline basins $\Delta S < 0$, that is, salinity decreases seaward from inside the basin.

11.2 Density-Gradient Reversals

In studying *reverse* or inverse basins, a crucial question is to determine how and when the basin can reverse. The answer to this question depends on the density gradient $\Delta\rho$ between the basin and the adjacent coastal ocean (Figure 11.1). Salinity may or may not control the density field. In many instances, it is necessary to consider the thermal contributions, that is, the relative contribution of temperature and salinity to density gradients (Section 2.4). In a typical estuarine basin (Figure 11.1a), the net volume flux is toward the coastal ocean when freshwater input, from river discharge or rain, exceeds evaporation. In a basin controlled by the temperature field and with warmer water in the basin, the net volume flux will be close to zero but the residual circulation shall be consistent with a typical estuary. On the other hand, when evaporation exceeds freshwater inputs (Figure 11.1b), the net volume flux will be directed into the basin. Once again, if the basin is controlled by cooling processes such as in high-latitude or in markedly seasonal areas with no negligible freshwater input, the net volume flux could be close to zero but the residual circulation will be consistent with an inverse system.

Scrutiny of the possible scenarios in which both water temperature and salinity gradients contribute to water density gradients in a basin can be effected via ΔT vs. ΔS diagrams (Figure 11.2). This approach is analogous to using *T*-*S* diagrams to identify *T* or *S* signals in water masses. In the ΔT vs. ΔS diagrams used here for illustration, reference values are 20°C for water temperature and 35 g/kg for salinity. Positive ΔT values (Figure 11.2a) denote that temperature increases from

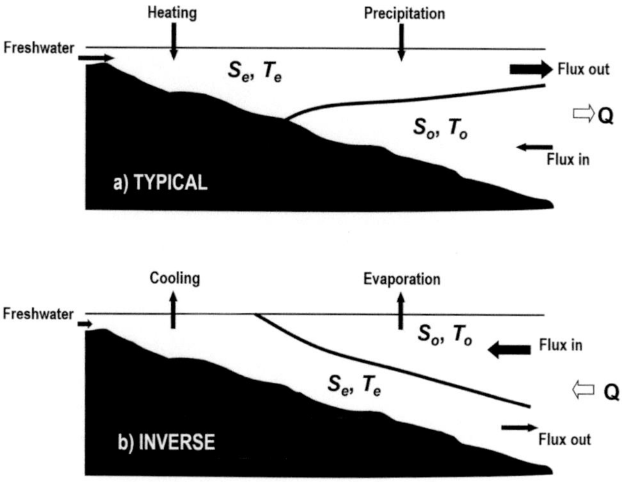

Figure 11.1 Scenarios expected for typical and inverse residual circulations in semienclosed basins with low discharge. In the typical situation, buoyancy inputs, from heat or freshwater, exceed evaporation losses and the net volume transport is out of the basin. The residual circulation is outflow at surface (or over shoals, as in Chapter 7) and inflow underneath (or in channel). In the inverse scenario, negative buoyancy from evaporation or cooling is larger than positive buoyancy and the net volume transport is into the basin. The residual circulation is opposite to the typical.

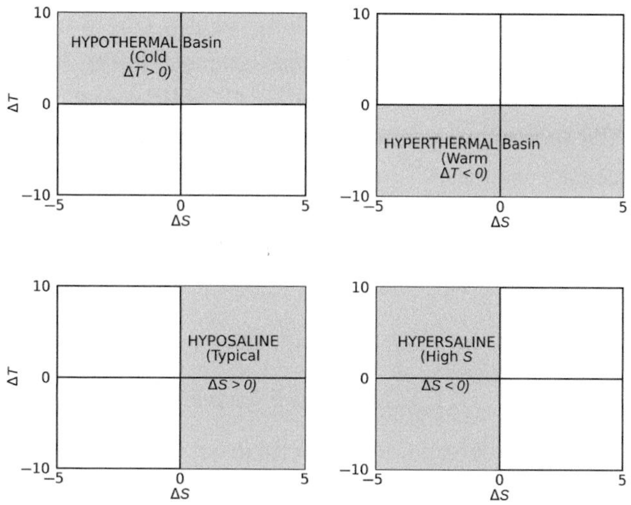

Figure 11.2 ΔT vs. ΔS diagram displaying four different types of conditions related to semienclosed basins: hypothermal, hyperthermal, hyposaline, and hypersaline.

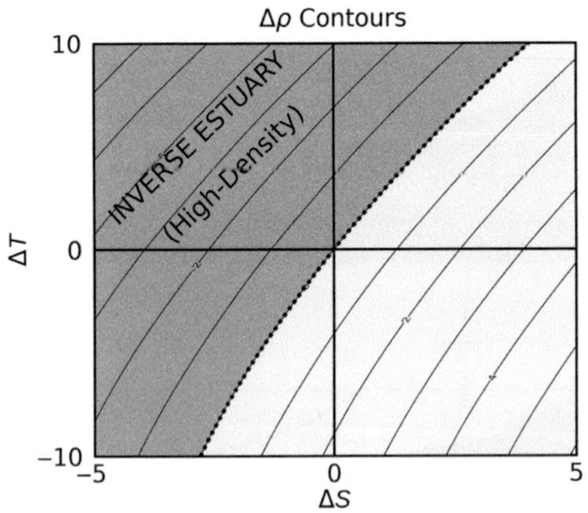

Figure 11.3 ΔT vs. ΔS diagram with $\Delta\rho$ isolines illustrating the regime ($\Delta\rho < 0$) for an inverse estuary (shaded region in the diagram). The dotted line denotes $\Delta\rho = 0$.

the basin to the ocean, that is, a cold or *hypothermal* basin. Similarly, negative ΔT values (Figure 11.2b) represent hyperthermal basins where water temperatures decrease seaward. With respect to salinity, positive ΔS values indicate basin waters that are relatively fresher than ocean waters, that is, *hyposaline* basins. Conversely, negative ΔS values occur in basins that have higher salinities than the coastal ocean as they are *hypersaline*.

In the same ΔT vs. ΔS diagram, we can draw isolines of density contrast $\Delta\rho$ between the basin and the ocean (Figure 11.3). Estimates of $\Delta\rho$ are obtained from the Thermodynamic Equation of Seawater (Section 2.3) relative to a reference water density with $S = 35$ g/kg and T of 20°C. Different reference values will affect the figure marginally. Positive contour values ($\Delta\rho > 0$) represent water densities increasing seaward, as in any typical estuary. In the ΔT vs. ΔS diagram, this is the portion to the right of the $\Delta\rho = 0$ isoline with white background (Figure 11.3). To the left and above the $\Delta\rho = 0$ isoline (shaded background in Figure 11.3), density decreases seaward ($\Delta\rho < 0$). In such parameter space, the basin will be *hyperpycnal* (density in the basin is greater than that in the ocean). If the main driver of the residual circulation in such basin is the density contrast, then the hyperpycnal basin will be inverse.

11.3 Thermal Estuaries

For the most part, we understand the basic hydrodynamics of regions with white and gray backgrounds in Figure 11.3. However, there are two regions in the ΔT vs. ΔS parameter space that represent fertile ground for research. One is the region where $\Delta\rho < 0$ (inverse) but $\Delta S > 0$ (hyposaline). In this region the basin is also

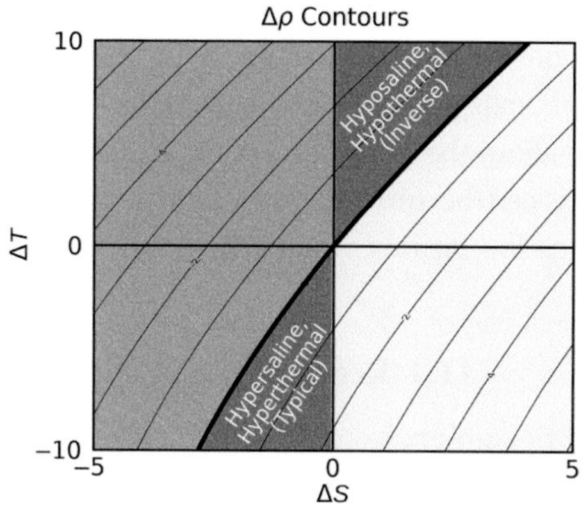

Figure 11.4 ΔT vs. ΔS diagram with two special cases in which the basin has a positive salinity gradient (hyposaline) but is inverse, and a negative salinity gradient (hypersaline) but is typical.

hypothermal (shaded with different tone on Figure 11.4). In such a situation, in which the salinity contrasts are relatively small and yet positive, the water density in the basin is determined by its relatively lower temperature than the adjacent ocean. Inverse circulation will develop in these hypothermal estuaries ($\Delta T > 0$). We expect to find this situation in high latitudes, where cooling reduces the water temperature of semienclosed basins more than the adjacent ocean. Coastal lagoons could also behave temporarily or transitionally as hypothermal estuaries in regions with marked cold air outbreaks and relatively weak tidal forcing, that is, with relatively low tidal Froude number (equation 8.17).

The second region in the ΔT vs. ΔS parameter space that merits attention is that in which $\Delta \rho > 0$ (typical) but $\Delta S < 0$ (hypersaline). Basins falling in this region are also hyperthermal (distinct shading in Figure 11.4), as $\Delta T < 0$. Systems in this space have received more attention but still provide an exciting research topic. Like in hypothermal estuaries, the basin–ocean salinity contrast is dynamically irrelevant in hyperthermal estuaries. In these hyperthermal systems the circulation is that of a typical estuary. Hyperthermal estuaries exist in Mediterranean-climate semienclosed basins also influenced by relatively cold coastal-ocean waters. In these regions, excess evaporation favors hypersaline conditions in the basin but not hyperpycnal. Cooler coastal waters adjacent to the basin may be provided by upwelling processes or by transport along the coast from a remote region. Hyperthermal estuaries may also occur in tropical latitudes with reduced or no freshwater influence where the basin becomes much warmer than the adjacent ocean. Hyperthermal estuaries exhibit typical residual circulation for periods determined by the prevalence of relatively cool coastal-ocean waters adjacent to the basin. During such periods, the tidal Froude number, like in hypothermal estuaries, should also be relatively small.

Both hypothermal (cold) and hyperthermal (warm) estuaries represent a class of semienclosed basins whose residual circulation is mostly driven by temperature gradients. As such, we can group these systems as thermal estuaries. Many questions are still open about the persistence of thermal estuaries and the temporal scales upon which they can be observable. Clearly, the development of thermal conditions will influence the basin's water renewal times and its water quality.

11.4 Regime Changes

There is another group of semienclosed basins, mostly in tropical and subtropical climates, that switch hydrodynamic behavior from typical to inverse. Loosely speaking, the switching in regimes can be referred to as *Stommel transitions* because of the proposed switching in the Atlantic Meridional Circulation regime originally advanced by Henry Stommel. Much work needs to be done on this topic. Most studies relate to hypersaline basins that show inverse residual circulation during a dry season and switch to hyposaline with typical residual circulation in the wet season. More efforts are required into placing basins that switch regime onto the ΔT vs. ΔS parameter space and follow their evolution at all temporal scales possible. Although the switch in circulation in these systems is expected to occur seasonally, the actual temporal scales of such a switching are unknown. Does the transition occur within a day? Does the regime change occur after several transitions back and forth, or after one switch?

The regime changes, just like proposed for *thermal* estuaries, should greatly affect water renewal times and water quality. We present two examples in subtropical and tropical regions that illustrate this point. The first example comes from a coastal lagoon in the Gulf of California, Guaymas Bay. During the dry season, centered in the Northern Hemisphere summer, the lagoon displays inverse estuarine residual flow in a transect at its entrance (Figure 11.5). The residual inflow at the surface, combined with residual outflow at depth, results from evaporation exceeding precipitation in those subtropical latitudes. During the winter season, the residual circulation resembles that of a typical estuary (Figure 6.5). In the winter situation, the outflow at surface and inflow at depth is actually driven by seaward winds. In other basins with marked wet to dry seasonality, this *typical* residual circulation could be driven by density gradients during the wet season (e.g., Figure 8.9).

The implications of a change in residual circulation regime are immediately evident in Guaymas Bay. During the inverse regime, untreated buoyant waters that are discharged onto the coastal ocean by a fish processing plant next to the lagoon's entrance end up inside the lagoon. Because these contaminated waters are trapped in the lagoon, the resulting water quality is rather poor in the dry season. The most obvious indicator of such water-quality conditions is the fetid sense

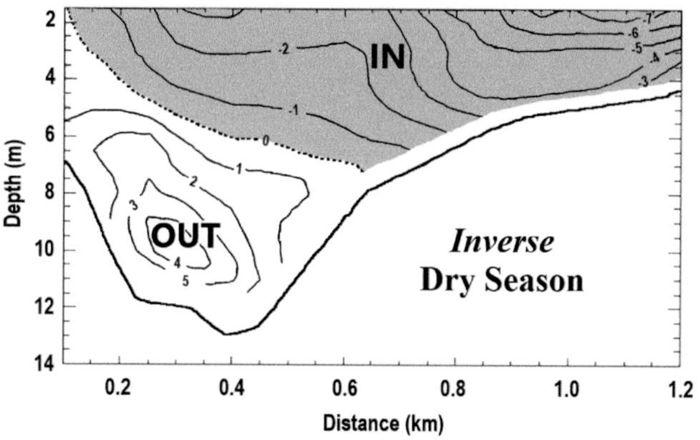

Figure 11.5 Residual circulation during the dry season at the entrance to a subtropical and semiarid lagoon, Guaymas Bay, in the Gulf of California. View is into the lagoon, with outflow contours in white background (toward viewer). The figure illustrates inverse circulation, in contrast to the typical circulation in the same lagoon during winter, which is portrayed in Figure 6.5.

around the coastal zone of the lagoon. In contrast, surface outflows drive the contaminated discharge offshore during winter. The ensuing water quality is much improved relative to the dry season, as indicated by the disappearance of unpleasant smells in the area.

The second example on regime switching comes from the tropical Gulf of Fonseca, a basin shared by El Salvador, Honduras, and Nicaragua, on the Pacific Coast of Central America. Also in a transect at the basin's entrance, the residual flow is typical (surface outflow vs. bottom inflow; or inflow in the deepest portion of the transect) in the wet season and inverse (bottom outflow, surface inflow) during the dry season. Similar to the findings in the semiarid bay, the water quality was apparently worse in the dry season because of evaporation-driven net volume inflow. This dry season situation could also exacerbate soil quality (becoming salty) and affect agriculture activities. On the other hand, water quality could deteriorate in the wet season because of an increase in land-derived pollutants carried to the estuary by enhanced river discharge and direct runoff from coastal areas.

11.5 Salt-Plug Estuaries

Another type of semienclosed basin found in subtropical and tropical regions may display typical and inverse circulation conditions at different portions of the basin (Figure 11.6). At the upstream reaches, density-driven circulation is typical, that is, outflow at surface and inflow underneath. Further seaward, the circulation may become inverse. This situation results in a maximum salinity area appearing inside the basin, that is, where the along-basin salinity gradient reverses sign. The

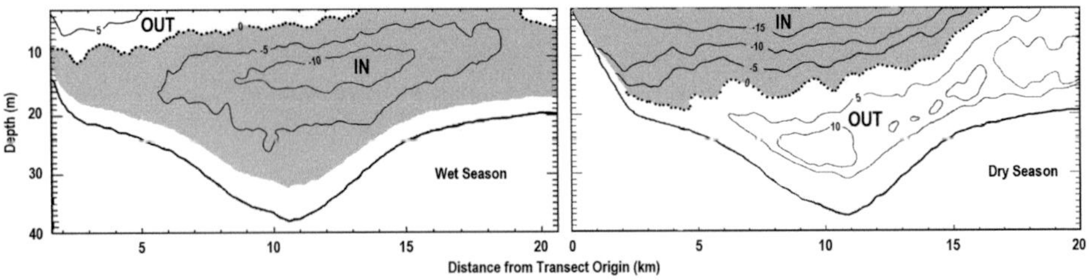

Figure 11.6 Residual circulation at the entrance to a tropical estuary, Gulf of Fonseca in Central America. Residual circulation switches from typical in the wet season to inverse in the dry season. View is seaward with flow away from the viewer in contours with white background. The transect spanned the middle half of the estuary entrance length. Most outflow in the wet season would appear to the right of the figure limits.

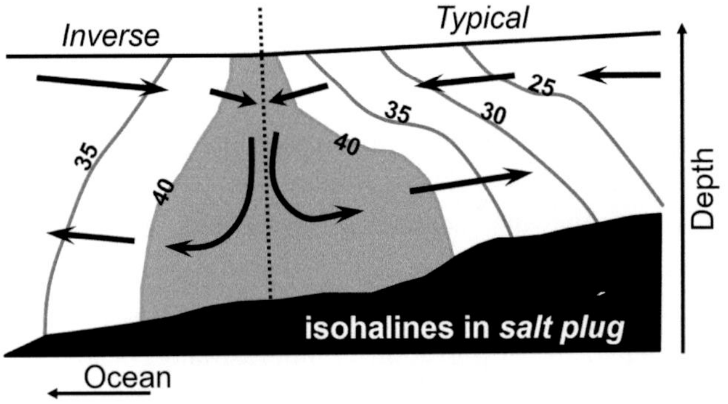

Figure 11.7 Schematic of an along-basin section in a salt-plug estuary. The diagram illustrates the maximum salinity portion inside the estuary. Landward of the salt plug, the residual circulation is typical whereas seaward of the plug, the circulation is inverse. The diagram also portrays a region of stagnant flow associated with the salt plug.

maximum salinity area is known as *salt plug* because it effectively blocks the exchange of waters between the lower and upper estuary.

Salt plugs have been described as the hypersaline portion surrounded by relatively lower salinity waters inside tropical basins. Examples of salt-plug estuaries have been documented in northern Australia and in arid northern Brazil, where evaporation volume rates are similar to river discharge values. However, salt plugs are not necessarily hypersaline. Regions of highest salinity inside a basin may arise in estuaries, including temperate basins, located downstream of another estuary with relatively higher buoyant discharge to the coastal ocean. This buoyant discharge can enter the estuary and promote a reversal in the along-basin salinity gradient in the lower estuary (Figure 11.7).

Analogous to inverse estuaries, salt-plug estuaries become susceptible to water-quality troubles. The salt-plug region becomes stagnant and acts as a barrier to water renewal in the basin. Problems of eutrophication, anoxia, and toxic algae blooms are likely to develop in these systems and require more attention in research activities. Processes associated with salt-plug estuaries should illuminate what might happen in areas where river discharge may increase as a consequence of climate change and thus influence neighboring bodies of water. Moreover, there is a series of questions and processes in semienclosed basins with low or no discharge that remain unresolved, providing a bountiful area of research.

11.6 Take-Home Message

Low-discharge semienclosed basins provide contrasting water exchange scenarios to those in typical estuaries. Such exchange scenarios are mostly determined by the density contrast between the basin and the adjacent ocean. The density contrast may be positive or negative and be determined by temperature gradients, by salinity gradients or by both. Still unresolved are a myriad of questions related to these systems. To address those questions, a first step is to examine the relative importance of temperature and salinity fields in the dynamics of the basin.

Additional Sources

Largier, J. (2010) Low-inflow estuaries: Hypersaline, inverse, and thermal scenarios. In *Contemporary Issues in Estuarine Physics*. Edited by A. Valle-Levinson, pp. 247–272. Cambridge: Cambridge University Press.

12

Classification of Semienclosed Basins, Based on Dynamics

This chapter is inspired by previous classifications of estuaries that have used prescribed physical parameters. Those classifications have been lucid and appropriate for diagnostic behaviors of density-driven residual circulation and stratification in estuaries (see Additional Sources for this chapter). However, those classifications omit tidally driven residual circulation in semienclosed basins as illustrated in Chapter 5. The proposed classification presented herein is based on tidally averaged, or residual, dynamics. The residual dynamics is treated from the perspective of forcing agents of residual flow, that is, tides or density gradients, balanced by modifiers, that is, frictional effects or Earth's rotation. The competition between forcing agents is determined by the nondimensional scaled, densimetric tidal Froude number, Fr_t, which depends on tidal current amplitude, reduced gravity, and water depth. In turn, the battle between modifiers, or counteracting forces, is established by the nondimensional Ekman number, *Ek*. This number depends on the eddy viscosity, the latitude, and the depth of the basin. The proposed classification identifies all types of estuaries, whose dynamics is influenced by density gradients, and also identifies coastal lagoons, tidal rivers, frictionless tidal basins, and systems that switch their dynamics from being tidally forced to being baroclinically forced. The proposed classification can also be regarded as a comparison of the strength of baroclinicity (in Fr_t) *vs.* the strength of mixing (in *Ek*) that ultimately determines the residual flows in semienclosed basins. Additional aspects of the classification consider systems forced by wind stress and/or baroclinicity, and basins forced by wind stress and/or tides.

12.1 Overview

Pioneering studies on estuarine hydrodynamics have inspired efforts to classify estuaries in order to diagnose or predict their function. Early propositions were based on salinity stratification and on the competition between river discharge and tidal forcing. Subsequent attempts suggested a stratification-circulation diagram that follows two nondimensional numbers: the circulation parameter and the stratification parameter. Missing from this scheme is an explicit inclusion of tidal forcing, which is typical of semienclosed basins.

Other propositions have included tidal forcing and density gradients in terms of different versions of Froude numbers (see Section 8.1.2 for one version). They also use a mixing number that compares the energetics of mixing vs. stratifying agents. These schemes can successfully identify different types of estuaries. Most classifications have the commonality of comparing stratifying tendencies vs. mixing agents. In such classifications, tides are considered a mixing agent but are ignored as potential drivers of residual flow (Chapter 5). To a large extent, classification schemes are determined by prescribed nondimensional numbers that disregard general influences of *Coriolis* acceleration and of tidally driven residual flows. This chapter offers a hydrodynamic classification that is general for semienclosed bodies of water. The classification is derived directly from a tidally averaged, along-basin momentum balance.

12.2 Fundamental Concepts

The proposed classification is anchored in the comparison between drivers of "residual" or tidally averaged circulation in semienclosed basins and actions that balance such drivers. This comparison between drivers and modifiers can be expressed in terms of a tidally averaged along-basin x momentum balance where local accelerations can be neglected (wind forcing is considered later):

$$g\frac{\partial \eta}{\partial x} + \frac{g}{\rho_0}\int_{-H}^{z}\frac{\partial \rho}{\partial x}dz + u\frac{\partial u}{\partial x} + v\frac{\partial u}{\partial y} = fv + \frac{\partial}{\partial z}\left[A_z\frac{\partial u}{\partial z}\right]. \tag{12.1}$$

Equation 12.1 is comparable to the first of equations 2.40 but with three small modifications: (i) no local accelerations, vertical velocities, or horizontal friction; (ii) all terms are averaged over one tidal cycle or longer periods; and (iii) the drivers or producers of tidally averaged flow are on the left-hand side while the modifiers or counterbalances are on the right-hand side. In equation 12.1 g is gravity's acceleration, η is water elevation, ρ_0 is a reference value for water density ρ, z is the vertical coordinate (positive upward), u and v are tidally evolving currents in the x and y directions, f is the *Coriolis* parameter, H is water column depth, and A_z is the kinematic eddy viscosity.

In the proposed classification, the main two drivers of residual flow are tidal forcing and pressure gradient from river influence (baroclinicity and surface slope); wind forcing is considered later. Therefore, the competition between drivers, which determines the main driver of residual flow, can be characterized as their ratio (Section 8.4 and equation 8.17).

$$Fr_t = \gamma \frac{U_0^2}{g'H}, \tag{12.2}$$

representing a scaled, densimetric Tidal Froude number Fr_t in which U_0 is the tidal current amplitude, H is water column depth, γ is the ratio between density-gradient length scale L_ρ and tidal length scale L_T (tidal excursion), and g' is reduced gravity ($g\Delta\rho/\rho_0$). In estuarine regions, $g\Delta\eta$ an $g'H$ reinforce the pressure gradient but $g'H$ is responsible for the gravitational circulation. Equation 12.2 implicitly incorporates the influence of river flow, by modifying g', as considered in other classifications.

The modifiers of residual flows, on the right hand-side of equation 12.1, are *Coriolis* acceleration and stress divergence (frictional effects). *Coriolis* accelerations may be scaled as $f\,U$, while the stress divergence may be scaled as

$$A_z \frac{U}{H^2}. \tag{12.3}$$

The competition between friction and *Coriolis* can be described with the ratio between equation (12.3) and the scaled *Coriolis* acceleration, through the vertical Ekman number (*Ek*, equation 7.40):

$$\frac{friction}{Coriolis} = \frac{A_z}{fH^2} = E_k. \tag{12.4}$$

This number may be regarded as a dynamical depth of a semienclosed basin and has been used to determine whether exchange flows in estuaries are vertically sheared or laterally sheared (Section 7.4). Although a given value of A_z is uncertain, we know that reasonable values fall in the range 10^{-2} to 10^{-4} m^2/s, or in even a narrower range. The important parameter is the nondimensional number (equation 12.4).

12.3 Classification Scheme

The proposed classification of the dynamics of semienclosed basins falls in a parameter space that considers Fr_t and *Ek*. The space of these two nondimensional numbers is formulated in a logarithmic scale. This approach stretches values that are <1 and contracts values >1. When the numerator of Fr_t or *Ek* is greater than the denominator (equations 12.2 and 12.4), the logarithm value is positive. Such value is zero when the denominator equals the numerator, and negative when the denominator dominates.

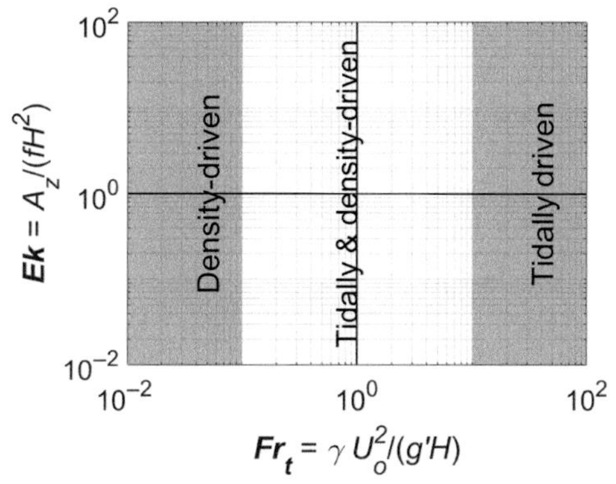

Figure 12.1 Parameter space in the proposed dynamics-based classification of semienclosed basins, focusing on the competition between tidal forcing and baroclinicity. Based on the scaled, densimetric tidal Froude number, tidally driven basins occur at $Fr_t > 10$, while density-driven basins appear at $Fr_t < 0.1$. At values $0.1 < Fr_t < 10$, the basins are expected to be influenced by both density gradients and tidal stresses.

12.3.1 Parameter 1: Tidal Forcing vs. Pressure Gradient

Consider first the two agents that are deemed drivers of the residual flow. This competition can be characterized in terms of Fr_t (Figure 12.1). Values lower than 0.1 indicate that the basin is essentially density-driven or forced by baroclinicity. Tidal forcing does not drive residual flows in this case, but it may modulate the residual flows by redistributing the baroclinic field of the pressure gradient. Values of $Fr_t = 0$ would indicate no tidal forcing and would fall well beyond to the left of the scale drawn in Figure 12.1. This situation represents flows driven by surface slopes, that is, a river (outside the scope here). On the other end, $Fr_t > 10$ indicates that residual flows are driven by tides (tidal residual flows – Chapter 5). Intermediate values of Fr_t identify systems where both tides and baroclinic pressure gradients can drive the residual flow. Thus, the horizontal axis in Figure 12.1 reveals the degree of river-induced baroclinicity in the basin. The values of 0.1 and 10 are arbitrary but indicate 10% or less influence of the other dynamic factor. The dynamic importance of baroclinicity decreases from left to right in the $Fr_t - Ek$ parameter space.

12.3.2 Parameter 2: Friction vs. Earth's Rotation

Focusing now on the *Ek* dependence (vertical axis on Figures 12.1 and 12.2), this number elucidates the dynamic depth of a system. Values of $Ek > 10$, that is, friction being 10 times, or more, greater than *Coriolis* accelerations, indicate highly frictional systems. Values of $Ek > 10$ mean that frictional or mixing effects

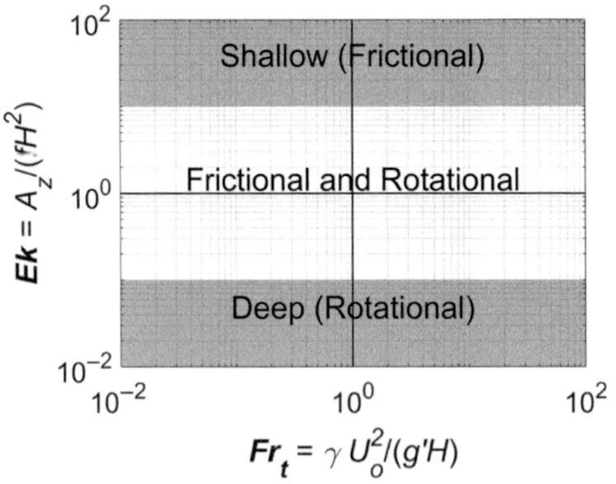

Figure 12.2 Parameter space in the proposed dynamics-based classification of semienclosed basins, focusing on the competition between *Coriolis* acceleration and friction. Based on the Ekman number, dynamically shallow or frictional basins occur at $Ek > 10$, while dynamically deep or rotational basins appear at $Ek < 0.1$. At values $0.1 < Ek < 10$, the basins are expected to be influenced by both rotation and friction. Vertical mixing should increase upward in the ordinate.

are felt throughout the entire water column, and *Coriolis* accelerations are irrelevant. Systems with these *Ek* values would be dynamically shallow and are illustrated by the upper shaded portion of Figure 12.2. Consider then the opposite situation in which *Coriolis* accelerations are 10 times greater than friction, that is, $Ek < 0.1$. This situation is shown as the lower shaded portion of Figure 12.2 and describes inviscid, or nearly inviscid, conditions where rotation dominates. Basins with $Ek < 0.1$ are considered dynamically deep. Thus, the vertical axis in Figure 12.2 is a proxy for mixing in the water column, with mixing decreasing from top to bottom. Overall, the $Fr_t - Ek$ parameter space identifies baroclinicity vs. mixing tendencies in the dynamics.

12.3.3 Partitioning in Terms of Fr_t and *Ek*

Baroclinically driven basins that are dynamically shallow will occupy the upper left quadrant of the parameter space (Figures 12.1 and 12.2). The dynamics of shallow, baroclinically driven systems can be simplified as the pressure gradient being balanced by friction. These dynamics exemplify a vast number of estuaries in temperate regions. They traditionally are known as "partially mixed" or "weakly stratified" or "partially stratified" estuaries (e.g., Figure 12.3). On the other extreme of dynamic depths, baroclinically driven basins that are dynamically deep will have negligible frictional effects and, in some cases, will be in geostrophic balance. These systems occupy the lower left quadrant of the diagram (Figure 12.3), are highly or "strongly stratified," and can include fjords. Between

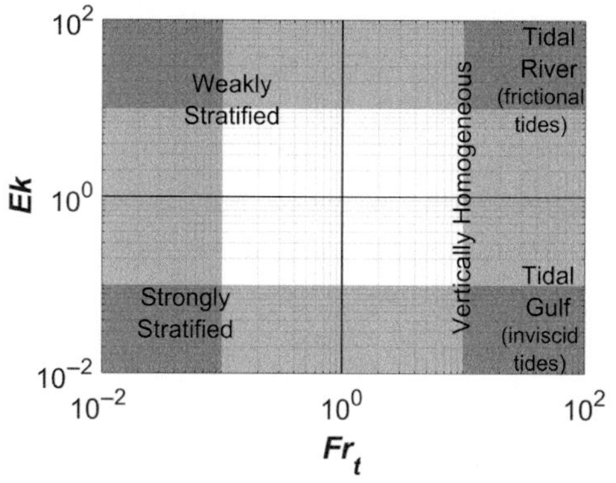

Figure 12.3 Different types of semienclosed basins forced by density gradients and tides, according to their location in the parameter space. Regions around $Ek \sim 1$ and $Fr_t \sim 1$ could switch regimes depending on spatial and temporal variations of forcing and bathymetry.

the dynamically shallow and dynamically deep extremes, there are systems that change from weakly stratified to strongly stratified, oscillating around $Ek = 1$ in the diagram (Figure 12.3). Switching from strongly stratified to weakly stratified or mixed can occur within one tidal cycle, that is, systems influenced markedly by tidal straining (Chapter 1); or can develop from neap to spring tides as stratification is modulated by tidal mixing; or change from dry to wet seasons as stratification is influenced by freshwater input or evaporation. Moreover, these changes can also be found instantaneously at different locations of the same basin because of differences in bathymetry and forcing. The great majority of studies in estuaries fall into this portion of the diagram ($Fr_t < 1$).

Tidally driven systems that are dynamically shallow will be found in the upper-right quadrant of the $Fr_t - Ek$ diagram (Figure 12.3). In the extreme right of the same upper-right quadrant, that is, $Fr_t > 10$ and $Ek > 10$, bottom stresses will be dynamically dominant. The dynamic balance will be between tidal stress and bottom friction. Such is the case of tidal rivers, shallow lagoons, or extremely energetic (tidal currents > 2 m/s) channels. In general, the residual flows in these systems are driven by frictional tides. In the lower-right quadrant of the $Fr_t - Ek$ diagram (Figure 12.3), frictional effects become negligible and tides are only influenced by Earth's rotation, analogous to Kelvin waves, that is, *tidal gulfs*.

12.3.4 Transitional Regimes

Two transitional regions may be relevant in the two right quadrants of Figure 12.3, that is, $Fr_t > 1$, where tidal forcing is the main driver of residual flows. One region spans horizontally around $Fr_t \sim 10$, in which the residual flow may have a very weak influence of baroclinicity, such as where the driving mechanism can be the covariance between eddy viscosity and vertical shear in tidal currents (Section 8.3). This region is labeled as "vertically homogeneous" on Figure 12.3 and may be spanned from spring to neap tides, or from wet to dry seasons depending on the buoyancy input and redistribution in the basin.

The second transitional region is oriented vertically around $Ek = 1$ in which the structure of tidally driven flows may change from neap or weak tides (dynamically deeper, $0.1 < Ek < 1$) to spring or strongest tides (dynamically shallower, $1 < Ek < 10$). This region can also change from annual maxima in tidal forcing to annual minima. The two types of transitions may also occur at different locations in a basin at a given time because of bathymetric H variations, resulting in different Ek and Fr_t.

12.3.5 Types of Semienclosed Basins according to Variables in Fr_t and Ek

The range of options in which any given basin may fall can be explored by assigning conceivable values to the six variables that anchor the proposed classification: (i) the eddy viscosity A_z; (ii) *Coriolis* parameter f; (iii) water depth H; (iv) tidal current amplitude U_0; (v) the scaling factor γ; and (vi) the density contrast along the estuary $\Delta\rho$. The different possibilities are outlined in Table 12.1. Particular values of the variables are nominal, but their dynamic significance resides in the nondimensional numbers Ek and Fr_t.

All combinations of variables on Table 12.1 are placed in the parameter space of Fr_t vs. Ek (Figure 12.4). Such placement shows distribution in different types of semienclosed basins. All values of $Fr_t < 1$ show estuarine types, density-driven basins, proposed in previous classifications. Fjord systems are density driven and

Table 12.1. *Parameters assigned to each one of the variables that enter the proposed classification in Figure 12.4**

Basin	A_z (m^2/s)	f (s^{-1})	H (m)	U_0 (m/s)	$\Delta\rho$ (kg/m^3)
Fjord	[0.001, 0.01]	10^{-4}	[100, 300]	[0.1, 1]	[1, 30]
Strongly stratified	[0.0005, 0.001]	[10^{-5}, 10^{-4}]	[10, 30]	[0.1, 0.5]	[10, 30]
Weakly stratified	[0.0007, 0.02]	[10^{-5}, 10^{-4}]	[5, 15]	[0.1, 1]	[20, 30]
Salt wedge	[0.0007, 0.001]	[10^{-5}, 10^{-4}]	[5, 15]	[0.3, 0.7]	[20, 30]
Frictional tides	[0.01, 0.02]	[10^{-5}, 10^{-4}]	[2, 10]	[0.2, 1]	[0.005, 0.04]
Frictionless tides	[0.001, 0.01]	[10^{-5}, 10^{-4}]	[100, 200]	[0.3, 1]	[10^{-4}, 10^{-3}]

* Numbers in brackets denote range of variation prescribed. The scaling factor γ in equation (4) is varied from 1 to 10.

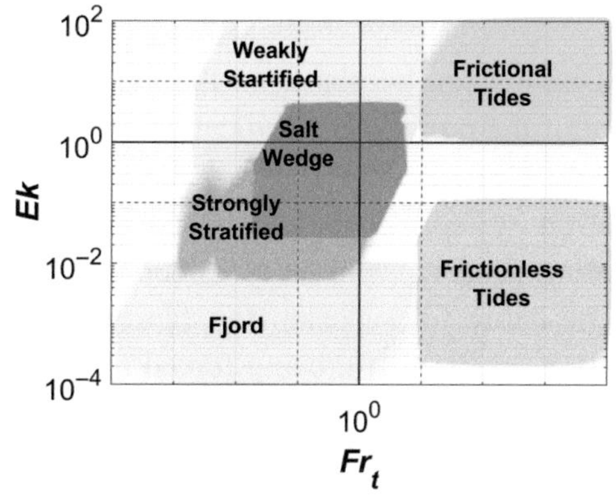

Figure 12.4 Types of semienclosed basins organized in the Fr_t vs. Ek parameter space.

deep ($Ek < 0.01$), although their residual circulation can also show influence from tides when $Fr_t > 0.1$. Strongly stratified basins are also dynamically deep ($Ek < 0.5$) and density driven ($Fr_t < 0.1$). There is overlap among fjords, strongly stratified estuaries, and salt-wedge estuaries. In fact, they are all stratified. Salt-wedge estuaries in the low end of Ek limiting these basins have permanent salt wedges. Toward the high end of Ek and Fr_t, within the salt-wedge limits on Figure 12.4, salt wedges most likely appear only in the flood stages of the tidal cycle (tidal salt wedge).

Toward the high end of Ek in the parameter space examined, remaining within low Fr_t values, the flow is still density driven but in dynamically shallow conditions. This is consistent with weakly stratified conditions that exhibit the classic estuarine exchange of outflow at surface and inflow underneath. But this combination of parameters could also apply to inverse estuaries if $\Delta\rho$ values have the opposite sign. On the high end of the Fr_t, as explained in Figure 12.3, the systems are vertically homogeneous and residual flows are only tidally driven.

Most relevant in this proposed classification are estuaries that fall in the square limited by $0.1 < Fr_t < 10$ and $0.1 < Ek < 10$. Falling in that region of the parameter space (Figure 12.4) means that the estuary's dynamics can either migrate horizontally around $Fr_t = 1$ or vertically around $Ek = 1$. The implication is that, for a horizontal excursion in the parameter space, the estuary's residual exchange flow can be forced at times by tides ($Fr_t > 1$) and at other instances by density gradients ($Fr_t < 1$). If there is also a vertical excursion around $Ek = 1$, the estuary's exchange flow shall switch from vertically sheared ($Ek < 1$) to laterally sheared ($Ek > 1$). It is possible that weakly stratified and tidal salt-wedge estuaries fall in this transitional parameter space (Figure 12.4).

12.3.6 Examples

Several semienclosed basins can be placed in the Fr_t vs. Ek parameter space to illustrate the relevance of the proposed classification. A total of seven systems are listed in Table 12.2, with the range of possible values of the five variables given in Table 12.1. These seven systems display some overlap in the parameter space (Figure 12.5), indicating dynamic similarity under certain tidal and buoyancy forcing conditions. The entire Chesapeake Bay can exhibit weakly and strongly stratified conditions ($0.01 < Ek < 10$), being forced mainly by density gradients. According to the proposed scheme, under certain circumstances of strong tidal and weak buoyancy forcing, tides may even influence the residual circulation. The Hudson River estuary is expected to display a subset of conditions shown by the Chesapeake Bay, that is, some regions of both estuaries can occasionally be dynamically similar. An equivalent dynamic behavior can be described for Mobile Bay, Alabama, where density gradients are expected to be even more influential

Table 12.2. *Parameters assigned to each one of the variables that enter the examples in Figure 12.5**

Basin	A_z (m²/s)	f (s⁻¹)	H (m)	U_O (m/s)	$\Delta\rho$ (kg/m³)
Chesapeake Bay	[0.002, 0.02]	0.87×10^{-4}	[5, 40]	[0.3, 1]	[15, 30]
Hudson River	[0.005, 0.02]	1×10^{-4}	[5, 20]	[0.4, 1]	[10, 30]
Tampa Bay	[0.005, 0.02]	0.68×10^{-4}	[3, 8]	[0.1, 1]	[5, 10]
Mobile Bay	[0.0001, 0.001]	0.73×10^{-4}	[4, 10]	[0.1, 0.5]	[20, 30]
Reloncavi Fjord	[0.001, 0.005]	1×10^{-4}	[50, 100]	[0.1, 0.3]	[20, 30]
Suwannee River	[0.001, 0.001]	0.70×10^{-4}	[1, 4]	[0.3, 0.5]	[0.5, 20]
Gulf of California	[0.001, 0.01]	$[0.6, 0.7]10^{-4}$	[100, 1000]	[0.5, 1.5]	[0.1, 1]

* Numbers in brackets denote range of variation prescribed. The scaling factor γ in equation (4) is varied from 1 to 10.

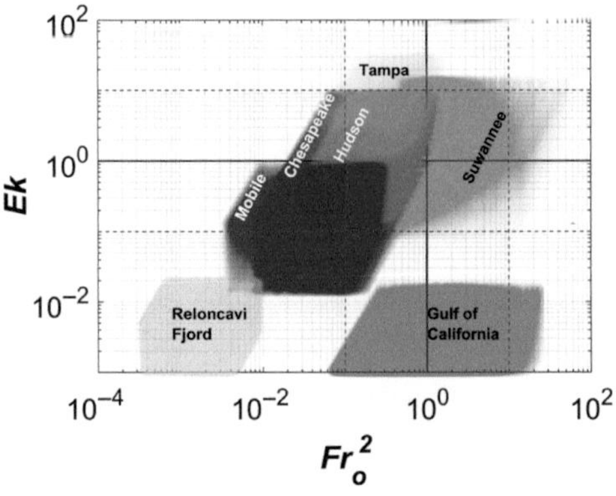

Figure 12.5 Specific examples drawn with the variables given in Table 12.2.

than in Chesapeake Bay and the Hudson River. Reloncavi Fjord in Chile is highly stratified and dominated by the density field. Tampa Bay, Florida, is weakly stratified and may be driven by both density gradients and tides. The Suwannee River estuary in Florida displays behaviors of a tidal salt wedge and a tidal river. The Gulf of California exhibits characteristics of an inviscid tidal basin with influence also from density gradients.

12.3.7 Other Special Cases

Several weakly stratified estuaries exemplify transitions in area of the $Fr_t = 1$ and $Ek = 1$ parameter space. Laguna San Ignacio, in the middle Pacific coast of the Baja California peninsula, features residual flows that are tidally driven and laterally sheared during spring tides. At this phase of the tidal fortnight, the estuary occupies the upper-right quadrant. During neap tides, the estuary's dynamical behavior changes to the lower-left quadrant as residual flows become density driven and vertically sheared. This is a dynamic behavior that also describes all possibilities for the Suwannee River in west Florida (Figure 12.5). Other weakly stratified examples with the same types of transitions are the Guadiana Estuary, separating southeastern Portugal from southwestern Spain, and the Mossoro Estuary in the dry tropical region of northeastern Brazil. The macrotidal Gironde Estuary, on the southwestern coast of France, can go from weakly stratified in spring tides to moderately stratified in neap tides. This change is accompanied by similar transitions in residual flows: from tidally driven and laterally sheared to density driven and vertically sheared.

Although there may be unpublished results of examples on transitions in the $Fr_t - Ek$ parameter space for tidal salt-wedge estuaries, like the Suwannee River, there may be several candidates in tropical regions. Unpublished data suggest that the tropical Magdalena River delta, flowing through the city of Barranquilla on the Colombian Caribbean coast, can go through these transitions. In the wet season, the delta is mostly a tidal river all the way downstream to the ocean (upper-right quadrant of Figures 12.4 and 12.5). In the dry season, the delta displays salt wedge during flood tides and the residual flow is mostly density driven (lower-left quadrant). Another similar unpublished example of a tidal salt-wedge system is the Alvarado Lagoon. This estuary is found in the Mexican state of Veracruz, on the western Gulf of Mexico.

12.4 Influence of Wind

Semienclosed basins are also forced by wind stress τ. This section proposes a classification of basins that are predominantly forced by wind stress. Two separate

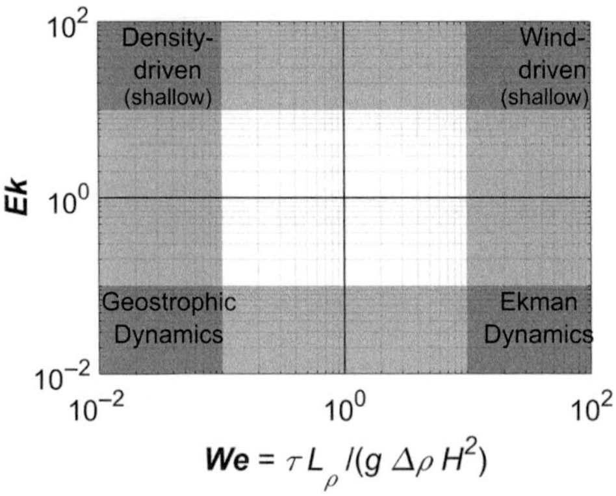

Figure 12.6 Different types of semienclosed basins forced by wind and density gradients, according to their location in the parameter space. Regions around $Ek \sim 1$ and $We \sim 1$ could switch regimes depending on spatial and temporal variations of forcing and bathymetry.

forcing agents are considered to compete with τ in this classification: density gradients and tidal stress. First considered is the competition between wind stress and density gradients. This competition may appear in micro-tidal estuaries and even in closed basins like lakes. The competition between τ and density gradients can be characterized with the Wedderburn number *We* (Equation 8.19):

$$We = \frac{\tau \, L_\rho}{g \Delta\rho H^2}. \tag{12.5}$$

Reactive (or balancing) forces to τ and density gradients can still be determined with *Ek* (equation 12.4). Basins with such dynamics are cast in the *We* vs. *Ek* parameter space (Figure 12.6). The left end of the diagram portrays basins with density-driven residual flows. The upper-left corner illustrates typical estuaries where density gradients are balanced by friction, that is, dynamically shallow density-driven flows. The lower-left corner of the *We* – *Ek* diagram represents density gradients balanced by *Coriolis*, that is, geostrophic dynamics, as in deep estuaries and density-driven lakes. On the right end of the diagram (Figure 12.6), the dynamics are dominated by wind stress. In the upper-right corner, wind stress is balanced by friction in such a way that winds will drive downwind near-surface flows and upwind near-bottom flows over relatively flat bathymetry (Chapter 6). Still in the upper-right corner (Figure 12.6), winds blowing over channel-shoal bathymetry will drive downwind flow over shoals and upwind flow in the channel (Chapter 6). In the lower-right corner, wind stress is balanced by *Coriolis* accelerations, that is, Ekman dynamics.

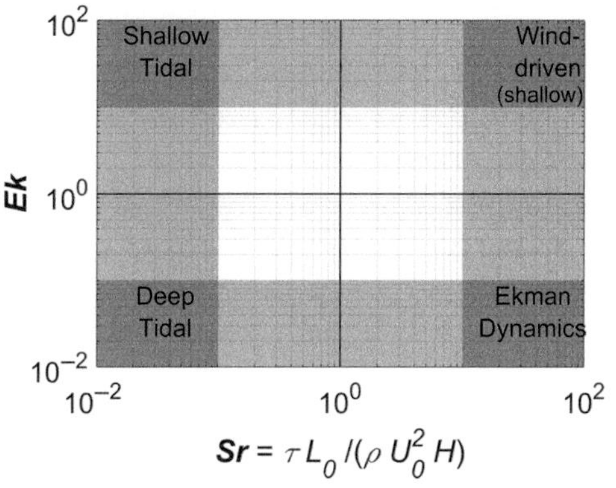

Figure 12.7 Different types of semienclosed basins forced by wind and tidal stresses, according to their location in the parameter space. Regions around $Ek \sim 1$ and $Sr \sim 1$ could switch regimes depending on spatial and temporal variations of forcing and bathymetry.

In the other possibility for wind stress, under no baroclinicity, the competition between τ and tidal stress can be assessed with the Stress number *Sr*:

$$Sr = \frac{\tau L_0}{\rho U_0^2 H}. \tag{8}$$

On the left end of the $Sr - Ek$ parameter space (Figure 12.7), tidal stresses dominate the residual flow. The upper-left corner illustrates tidal residual flow in shallow basins while the lower-left corner portrays deep basins with tidal residual flows. The right end of the $Sr - Ek$ parameter space (Figure 12.7) describes basins where the residual flows are wind driven. This portion of the diagram is equivalent to the same diagram portion of Figure 12.6. The upper-right corner represents flows driven by the wind in dynamically shallow basins while the lower-right corner illustrates Ekman dynamics. This part of the classification applies to semienclosed basins where density gradients are negligible, such as coastal lagoons and relatively homogeneous bays and gulfs.

12.5 Take-Home Message

The proposed classification of semienclosed basins follows a tidally averaged momentum balance that compares drivers and modifiers of residual flow. Residual flow drivers are characterized by the nondimensional Fr_t, while balancing forces (per unit mass) are typified by another nondimensional number, *Ek*. Thus, the classification is contained in the $Fr_t - Ek$ parameter space and considers estuaries, tidal rivers, vertically homogeneous lagoons, and frictionless tidal basins. The scheme can be regarded as the baroclinic tendencies shown by any basin, presented in the abscissa, vs. the mixing tendencies, as represented in the ordinate.

Also considered are situations in which wind stress competes with (a) density gradients, as represented by *We* and (b) tidal stress, as characterized by *Sr*. This additional approach allows a dynamic description of basins where the dynamics are purely frictional, non-frictional (geostrophic), and Ekman-type. This classification includes semienclosed basins beyond estuaries.

Additional Sources

Cameron, W.M., and D.W. Pritchard (1963) Estuaries. In *The Sea*. Edited by M.N. Hill, pp. 306–324. New York: Interscience.

Cheng, P., H.E. De Swart, and A. Valle-Levinson (2013) Role of asymmetric tidal mixing in the subtidal dynamics of narrow estuaries. *J. Geophys. Res.: Oceans* 118: 2623–2639.

Geyer, W.R. (2010) Estuarine salinity structure and circulation. In *Contemporary Issues in Estuarine Physics*. Edited by A. Valle-Levinson, pp. 12–26. Cambridge: Cambridge University Press.

Geyer, W.R., and P. MacCready (2014) The estuarine circulation. *Ann. Rev. Fluid Mech.* 46: 175–197.

Guha, A., and G.A. Lawrence (2013) Estuary classification revisited. *J. Phys. Ocean.* 43 (8): 1566–1571.

Hansen, D.V., and M. Rattray Jr. (1965) Gravitational circulation in straits and estuaries. *J. Mar. Res.* 23: 104–122.

Hansen, D.V., and M. Rattray Jr. (1966) New dimensions in estuary classification 1. *Limnol. Oceanogr.* 11(3): 319–326.

Pritchard, D.W. (1956) The dynamic structure of a coastal plain estuary. *J. Mar. Res.* 15(1): 33–42.

Stommel, H., and H.G. Farmer (1952) Abrupt change in width in two-layer open channel flow. *J. Mar. Res.* 11: 205–214.

Valle-Levinson, A. (2021) Dynamics-based classification of semienclosed basins. *Reg. Stud. Mar. Sci.* 46: 101866.

Index

Printed in the United States
by Baker & Taylor Publisher Services